Assessing Urban Governance

The Case of Water Service Co-production in Venezuela

P.I.E. Peter Lang
Bruxelles · Bern · Berlin · Frankfurt am Main · New York · Oxford · Wien

Luisa Moretto

Assessing Urban Governance

The Case of Water Service Co-production in Venezuela

« Public Action »
vol. 12

With the financial support of the ULB Faculty of Architecture.

This publication has been peer-reviewed.

Éditions scientifiques internationales
Brussels, 2014
1 avenue Maurice, B-1050 Brussels, Belgium
pie@peterlang.com; www.peterlang.com

ISSN 1783-6077
ISBN 978-2-87574-227-8
eISBN 978-3-0352-6494-4
D/2014/5678/99

Printed in Germany

Cip available from the British Library, UK and from the Library of Congress, USA.

"Die Deutsche National Bibliothek" lists this publication in the "Deutsche Nationalbibliografie"; detailed bibliographic data is available on the Internet at http://dnb.de.

To Pascal

Table of Contents

Acknowledgments

First of all, I am genuinely indebted to Marcello Balbo, professor at the University IUAV of Venice, Italy, who has always been available to meet up and discuss my research, as well as read and comment on the chapters that made up the previous version of this book. I am also very grateful to Julio Davila, professor at the Developing Planning Unit, University College London, for his valuable help in conducting my research and for sharing his experience with me. I am also grateful to two professors I met at the CENDES, Centro de Estudios del Desarrollo, Universidad Central de Venezuela: Miguel Lacabana, who illustrated the findings of his research, orientated my work on the ground and provided the indispensable community leader contacts through which to undertake the fieldwork; and Marianela Carillo for physically introducing me in the communities studied. I would also like to express my gratitude to Shipra Narang, Vice President at ISOCARP, the International Society of City and Regional Planners, and to Sylvy Jaglin, professor at the Université Paris-Est Marne-la-Vallée, for their precious comments on the first version of this research, which helped me immensely in finalising my work.

It would have not been possible to conduct this research without the welcome, availability and support of the community members in the two Venezuelan settlements where the fieldwork was conducted. In particular, Elena, Yunilde and Gladys, who let me stay in their homes and always took care of me; they spent their time facilitating this research and providing the best conditions under which to conduct the interviews. The memory of their enthusiasm and courage in working to improve their settlement's living conditions was a wonderful incentive to complete this book. A special thanks to Norieli too, for keeping me company with her drawings during the transcription of the interviews.

Thanks also to Pascal, for producing all the graphics in this book with incredible patience, despite my never-ending remarks. Finally, thanks to Jean-Louis Genard for the opportunity to publish my work in this PIE Peter Lang collection.

Introduction

This book presents an exploration of the complex and uncertain domain of urban governance and sustainable water supplies for the benefit of the urban poor, with a precise focus on developing countries. Ensuring and governing sustainable access to basic services, as well as the wider urban development in the cities of the South, is still an open and debated issue. Different policies and strategies have been proposed over time in order to guarantee basic services for the urban poor. Generally, they have rested on the public service model, on private interventions and management techniques, and on multiple kinds of public-private partnerships – up to the current focus on, and confidence in, various forms of governance systems, seen as a way to bridge the gaps between the wide variety of service provision models. Since the 1990s, the international community has increasingly seen governance as the most promising development mechanism to also ensure sustainable access to basic services at the city level, although, as has been recurrently noticed, "its versatility means that it continues to mean different things to different authors" (Hydén, 2011: 5). Most probably due to the vagueness of the concept, governance presents a gap between rhetoric and reality not only in its definition and operation: what (urban) governance is, how it really works in practice and how it could be measured, distinguishing formal procedures and prescriptions from local actions and practices, remain open questions.

Governance, good governance and urban governance

Despite some researchers raising concerns about the prospect of governance becoming the "development fad" of the 1990s (Halfani *et al.*, 1995: 108), it is clear at the present time that this term has entered into the development agenda of all donors as a common element of aid-speak. Since the beginning of the 1990s, the term "governance" has been frequently recognised as a new buzzword used by the international aid community to improve development assistance (Boeninger, 1991; Paproski, 1993). Of all the many definitions considered, the one cited below highlights two fundamental governance connotations that enjoy a large consensus: first, governance is broader than government and, second, it involves all civil society stakeholders[1]: "Governance, as distinct from government,

1 Without entering the broad debate regarding the notion of "civil society" (Mitlin, 1999; Lewis, 2000; Stren, 2000 for some relevant references), it is sufficient to say that civil society is considered here in a broad sense such as "community-based and grass-root

refers to the *relationship* between civil society and the state, between rulers and the ruled, the government and the governed" (Halfani *et al.*, 1995: 108). Within this definition, governance relationships have been described as "joint action" (Rakodi, 2003: 524), "governing interactions" (Kooiman, 2003: 5), "negotiation mechanisms" (Garcia, 2006: 745), and "interact[ions] with a view to policy making (Hydén, 2011: 251), just to cite some examples.

In the context of development assistance, the concept of governance has often been replaced with that of *good governance*"[2]. Good governance has represented a strategic key issue in achieving development goals and has even been seen as a reform objective in itself (Degnbol-Martinussen and Engberg-Pedersen, 2003; Stren, 2012). Above all, however, it has assumed a deep and specific normative connotation for donor agencies that has, in turn, resulted in a clear conditionality for recipient countries (Harpham and Boateng, 1997; Devas, 1999; Castro, 2005; Hydén, 2011). According to Leftwich (1994: 371), good governance includes three main normative levels: systemic ("the concept of governance is wider than that of government" and "refers to a system of political and socio-economic relations"), political (refers to a multi-party representative democracy) and administrative ("an efficient, independent, accountable and open public service"). Nevertheless, the political and administrative means by which these norms could be put into practice have been frequently controversial, relying often on external donor pressure towards liberal democratisation and privatisation (Nunan and Satterthwaite, 2001). As a result, the normative identification of good governance characteristics is very often followed, in literature, by concrete descriptions of how good governance relationships and responsibilities should be operationalized (Rakodi, 2003). This is easily recognisable at the urban level.

Urban governance increasingly represents the local level of application of good governance prescriptions (normative as well as practical) in developing countries (Lombard, 2013). The interest in cities lies undoubtedly in the recognition of rapid urban growth and in the resulting fundamental role of the urban areas for development. It is within cities that productivity and social welfare flourish, making them one of

organisations, NGOs, trade unions, religious organisations and business, both formal and informal, alongside the various branches of governmental agencies, both national and local" (Amis *et al.*, 2001: 4-5).

2 According to Letftwich (1994: 370, "[t]he first official appearance of the contemporary notion of good governance came in the 1989 World Bank report on Africa, which argued that 'Underlying the litany of Africa's development problems is a crisis of governance', by which was meant 'the exercise of political power to manage a notion's affairs' (World Bank, 1989: 60)".

the privileged places where aid assistance is concentrated in order to achieve the broader goal of poverty reduction but also to strengthen their role in the global market (Osmont, 1995, 2002; Balbo, 2002b). Likewise, the city is the place where civil society mobilises and organises itself to cope with service shortcomings (see for instance Halfani *et al.*, 1995; Mayer and Künkel, 2012) and where in turn, profiting from an active civil society, governance relationships can better develop. In particular, giving emphasis to the principle of an improved equity, the involvement of communities – and especially low-income communities – turns out to be a promising alternative to the dominant hierarchical and capitalistic organisations. Basic services become a shared responsibility between the numerous urban stakeholders, which have to agree on and contribute both to the design and delivery of city services. Urban governance is supposed to provide the arena for urban actors – including those belonging to the community sector, notably through the notion of civil society – to be democratically represented in the institutional decision-making process and to directly play a part in service delivery through their concrete and innovative contributions. Nevertheless, the potential of the very rationale behind such kind of community participation to challenge the top-down governance approach – whether state – or market-driven – is often placed under discussion.

Local, informal, needs-driven, practice-based governance relationships and arrangements

Different types of governance arrangements and relationships can also unfold at the local level, according to a process based on the genuine involvement of communities as the central actors in the development process. In developing countries, when the public or private sectors fail to guarantee formal provisions, the urban poor rely on a broad variety of different practices to gain access to basic services and drinking water provisions. These systems are generally developed at the very local level, based on a wide spectrum of informal – and sometimes illegal – arrangements (belonging to market mechanisms or solidarity networks) and rooted in traditional or consolidated practices. They are usually "needs-driven", in contrast with the more formal and "policy-driven" mechanisms (Allen *et al.*, 2006a) supported by institutional bodies or development agencies.

Such informal survival practices for obtaining drinking water are often overlooked or even opposed. Nevertheless, evidence from research carried out in the last two decades[3] has shown that, when local institutions

[3] See for instance the following research works: Amis *et al.*, 2001; Jaglin, 2001; Nickson and Franceys, 2003; Batley and Moran, 2004; Allen *et al.*, 2004, 2006b; Satterthwaite,

recognise and support these needs-driven practices, new forms of collective arrangements originate between low-income communities and the other actors involved in the water delivery process. This is the way in which urban governance works in practice, at a very local- and practical-based level. Several positive outcomes and synergies can arise from these types of governance arrangements. They range from the improved quality of water provisions to a more efficient use of financial and natural resources; from a broader recognition of the urban poor as citizens entitled to numerous rights – urban services, land, housing, etc. – and their involvement in the decision-making process.

Many critics agree that these cooperative actions and arrangements need to be supported and included in the broader discourse regarding urban governance, as alternative governance systems to achieve structural improvements in water supplies. To this end, recent research has been focused, on the one hand, on a better understanding of the rationale and rules that characterise the informal and locally-based practices and on the potentialities that they carry. On the other hand, it has investigated how the more political and institutional processes can – formally or informally – support and collaborate with the wide variety of urban service providers, exploring ways in which to articulate informal and local practices to the more formal system of distribution, producing synergies rather than negative effects (see for instance Allen *et al.*, 2006a; Phumpiu and Gustafsson, 2008; Olivier de Sardan, 2009; Batley and Mcloughlin, 2010; Allen, 2010, 2012; Booth, 2011; Gaventa and Barrett, 2010; Wild *et al.*, 2012).

The case of technical water committees in the Caracas Metropolitan Region

Local collective actions for ensuring access to water provisions to low-income communities in Venezuela seem to fall into the above framework. They are considered alternative modes of governance coordination since community organisations play a fundamental role in shaping innovative "governance practices" where new relationships unfold amongst the different actors involved in the service delivery process (Lacabana, 2003; Cariola and Lacabana, 2003, 2005b; Lacabana *et al.*, 2004; Allen *et al.*, 2006a, 2006b; UN-HABITAT 2006a, Moretto, 2010; Allen, 2010, 2012; Aubriot and Moretto, 2013; McMillan *et al.*, 2014).

2005; Phumpiu and Gustafsson, 2008; Marin, 2009; Batley and Mcloughlin, 2010; Moretto, 2010; Wild *et al.*, 2012; Lombard, 2013.

Current Venezuelan systems of accessing water service provisions – especially for low-income residents – are based on both formal and informal practices and arrangements. In most Venezuelan municipalities, formal practices are represented by piped networks, public or private household distribution through licensed tankers, wells or bore-wells. Informal practices include illegal connections to the main public pipes or private distribution through non-licensed tankers[4]. For a little over a decade, these illegal water provision networks have gone unopposed. On the contrary, we have witnessed an explicit institutional effort to better organise and further develop these kinds of connections. The 1999 Constitution and the 2001 Drinking Water and Sanitation Service Act have been milestones in a new normative framework, aimed at regularising and institutionalising these needs-driven and informal practices.

At the core of this new tentative governance practice in Venezuela are the "technical water committees" – *Mesas Tecnicas de Agua* (MTAs) – which represent a new and fundamental community actor. An MTA is a community-based organisation, built for the purpose of channelling community participation in the decision-making process and carrying out the physical improvements to service delivery. Governance arrangements develop around MTAs as a result of agreements, collaboration and cooperation with the other main stakeholders in the drinking water provision process – notably, the water companies, the municipalities and all of the community members.

This type of governance practice is here presented as a form of "institutionalised co-production" (Joshi and Moore, 2004) where citizens directly and regularly interact with the more institutional actors in order to bring about structural improvements. Venezuelan current political orientations are indubitably a major factor in supporting the development of such co-production. The government's efforts towards looking for alternatives to capitalistic regimes are unquestionably a priority. If "Mr Chavez's fame stems, too [besides the opposition to George Bush], from his oil-backed claims to be leading a new revolutionary project of "21st-century socialism" that tears up the nostrums of economic liberalism"[5], current president Nicolas Maduro seems clearly to be pursuing Chavez politics. Nevertheless, whether this kind of institutionalised co-production could be considered as a genuine alternative governance mode or as still belonging to a state-centric approach needs to be further investigated.

---

4 As pointed out by Cariola and Lacabana (2005b), "informal" connections to the public piped network can be defined as "illegal" practices, which have been permitted and sometimes supported by the water companies.

5 *The Economist*, August 4th 2007, p. 70.

Investigating urban governance for better water provisions

Likewise with governance, urban governance is a multifaceted concept. It presents a formal, institutional, theoretical and normative aspect, broadly analysed and debated by the international community and unfailingly included in donors' development agendas. At the same time, this notion also displays a more informal, local, community-based feature, which describes how governance works in practice, beyond top-down and external development policies and strategies. Two different assessment approaches seem thus necessary in order to investigate the two urban governance facets. One approach should focus on quantitative and factual data and information, collected at the city level with the support of municipalities and local governments. The other should rely on the information related to local and informal arrangements for better services, which is, by its very nature, not systematically collected and organised into administrative databases. When urban governance for better provisions is appraised, a gap can clearly occur between reality and rhetoric, which is between the practice-based, mainly informal, local governance processes for providing improved urban services, and the formal governance prescriptions and mechanisms for service delivery within the institutional sphere.

As a result, assessing the extent and the quality of urban governance for better water provisions faces major difficulties – if not limitations – since both these different perspectives need to be recognised, distinguished, understood and correctly evaluated. They also need to be compared and mutually verified as they both have the same goal: to improve urban poor living conditions. To this end, this book aims at exploring two interlinked questions: first, are current urban governance assessments – mainly proposed by donor agencies – applicable to local water service provisions in the global South? Second, are they able to look into the two-fold dimension of governance to appreciate its formal architecture at the municipal level and also the nature of informal and locally-specific governance arrangements aimed at improving access to basic services? At the basis of these questions lies the idea that governance cannot belong exclusively to government hierarchies or market rules as modes of coordination. On the contrary, successful governance to achieve sustainable urban development and access to urban services rests upon the principles of inclusiveness and citizenship. Assessing urban governance means thus finding out whether and to what extent all the actors are involved in and contribute to the service delivery process, through an organisational system embracing both formal institutional processes and informal practices, and both institutional and political mechanisms and community organisations. In the case of Venezuela, it means assessing the existent degree of involvement and inclusion of all of the community members in the service co-production between the water committees and the water company, as well as the residents' perception of their inclusion into these governance arrangements.

Amongst the few instruments elaborated by the main international aid agencies to assess urban governance, this book focuses on the UN-HABITAT Urban Governance Index (UGI) to investigate the new governance arrangements to co-produce water services in Venezuela. The UGI is a tool elaborated within the UN-HABITAT Global Campaign on Urban Governance with the overall purpose of making a contribution to the eradication of poverty, by strengthening and improving urban governance. This instrument has been developed "in order to enable cities to objectively measure the quality of local governance" (Narang, 2005: 1) through a core set of quantitative indicators. The initial hypotheses in the application of this index are twofold, and will be only partially verified as presented in chapter 7. First, the UGI is a valuable instrument to investigate governance mechanisms as its selection of indicators has been carried out according to the principle of inclusiveness and it can provide a comprehensive framework for exploring and assessing the more local and informal aspects of local governance. Second, governance mechanisms built around the MTAs equitably and effectively involve all of the community residents in both the decisions involving, and contributions made to, the provision of water supply, together with all of the actors involved in the service delivery.

Overview of the book

The book comprises seven chapters and three annexes. The first chapter reviews the relevance that urban (water) services have had in the evolving significance of urban governance in developing countries. It focuses on the changing relationships between urban governance and service delivery while outlining, on the one hand, the different public service reforms promoted through the neo-liberal agenda of structural adjustment and the values and priorities underlying them and, on the other, the role of informal water providers in guaranteeing drinking water access for the marginalised urban poor.

Following an introduction to urban governance assessments, chapter 2 focuses on the methodological and conceptual frameworks concerning the analysis process employed in the application of the Urban Governance Index. It identifies and describes the UGI key assumptions and principles, as well as how its theoretical framework has been translated into qualitative analysis at the settlement level. Relying on the notion of "institutionalised co-production" (Joshi and Moore, 2004), chapter 3 presents the main actors involved in the water delivery process in the

Caracas Metropolitan Region, and the Tuy Valleys in particular, and the alternative governance arrangements they developed. This section is introduced by an overview of the current urban water supply regulation and organisation in Venezuela.

Chapter 4 provides a detailed description of the two analysed settlements regarding their history, tenure situation, the composition of the MTAs and their relationships with local, regional and national institutional bodies. The second part of the chapter is dedicated to presenting the way in which UGI principles and indicators have been used to conceive a multilevel mixed model design to assess urban governance for water service delivery in Venezuela. It also includes methodological information about the UGI's adaptation to the specific case study context, the data collection questions and hypothesis related to the case study, the sample choice and the sources of evidence applied in the different phases of the study. Explanations are also given of how the fieldwork has been carried out in both the municipalities and low-income communities of the Caracas Metropolitan Region in order to test the research hypotheses.

Chapters 5 and 6 present an analysis of the governance mechanisms developed in the two units of analysis of the research, that is, the municipal and community levels. This investigation is conducted relying on the four core UGI principles (effectiveness, equity, participation, accountability) and the resulting eight sub-criteria identified as the common research framework between the two levels of analysis. Chapter 5 aims at exploring and understanding the extent and the quality of governance at the municipal level of analysis in improving water supplies especially for low-income communities. Two Venezuelan municipalities represent the units of analysis in the first stage, as the two low-income communities chosen for this research belong to these two municipalities. Quantitative data collected through the calculation of the twenty-five indicators belonging to the Urban Governance Index are used to explore local governance in these municipalities. Through a "qualitising process", this chapter provides an explanation of the significance of UGI indicator data for the two selected municipalities and verifies whether it is possible to confirm a Venezuelan general trend regarding urban governance at the municipal level, always with a special focus on water provisions. It also highlights further aspects of local governance in Venezuela that help in gauging the appropriateness of the simple UGI indicators for assessing governance at the local level.

Chapter 6 presents the results of the qualitative analysis carried out in the two low-income settlements of the Caracas Metropolitan Region selected for this study. This chapter aims to assess to what degree the two settlements' residents were included and involved in the governance arrangements built around the water boards. Results rely on the evidence collected from the four principal groups of stakeholders involved in these governance arrangements (community members, MTA leaders, municipal public servants and state-owned water company officials) through a variety of research techniques, which have been systematically triangulated (interviews, documentary sources, participant and direct observation, focus groups, physical artefacts).

In the concluding chapter 7, meta-inferences from the two levels and units of analysis are worked out according to a narrative integration and comparison of the "qualitised" inferences presented in chapter 5 and the qualitative findings provided in chapter 6. The integration between these two orders of inferences aims at understanding and comparing the entire formal, more institutional dimension of urban governance at the municipal level, as well as the more informal, local and community-based aspects. The chapter concludes by evaluating the UGI's applicability to the specific problem of water service delivery, and presenting policy implications and possible inputs for future urban governance assessments.

Finally, three annexes complete the thesis. The first delivers the complete list of interviewees. The second presents the responses from the community members to the interviews conducted within the two samples. Annex 3 provides a detailed calculation of the twenty-five Urban Governance Indicators, adapted for the purpose of this research, in the two analysed municipalities.

Chapter 1

Urban Governance and Water Supply Systems

1.1. Urban Governance and Neoliberalism

Ronald Coase, in his article *The Nature of the Firm* (1937), is recognised as the first exponent of the neo-liberal governance concept. He used this term to describe the relationships of co-ordination and co-operation inside an enterprise to reduce the transaction costs generated by the market. In the 1970s and 1980s, this concept was rediscovered by some economists and termed "corporate governance" or "governance structures" to simplify and rationalise the organisation of enterprises following the principles of economic efficiency and managerial effectiveness (see for instance Brown, 2001; Osmont, 2002; Miegeville *et al.*, 2003; Maloutas and Malouta, 2004).

Born in the context of the efficient management of enterprises, the concept of governance has been transferred to the public sphere and tasked with coping with the increasing reduction of financial resources being experienced above all by local and urban administrations (Brown, 2001). This process took place during the 1980s in the cities of both developed and developing countries. Borrowing this concept from the managerial field, municipal administrations have begun to manage cities focusing on productivity rather than on their welfare in terms of housing, infrastructure and basic needs (Halfani *et al.*, 1995; Harvey, 2005; Guarneroz-Meza and Geddes, 2010; Didier *et al.*, 2012). Cities have thus increasingly represented a strategic actor from the perspective of a neoliberal type of development, and urban governance the process to guaranteeing neoliberalism in the urban sphere[1].

Neoliberalism and the city

Sketching the principal features of the shift from liberalism to neoliberalism, Jessop (2002: 454-455) distinguishes the main economic, political and social measures that underpin the neoliberal project:

[1] Major contributions to this research field are provided by: Burgess *et al.*, 1997; Osmont, 1999, 2002; Brown, 2001; Jessop 2002; Miraftab, 2004; Harvey, 2005, 2012; Leitner *et al.*, 2007; Guarneroz-Meza and Geddes, 2010; Theodore *et al.*, 2011; Künkel and Mayer, 2012; Peck *et al.*, 2013.

> As a new economic project oriented to new conditions, neoliberalism calls for: the liberalization and deregulation of economic transaction not only within national borders but also – and more importantly – *across* these borders; the privatization of state-owned enterprises and state-provided services; the use of market proxies in the residual public sector; and the treatment of public welfare spending as a cost of international production, rather than as a source of domestic demand. As a political project, it seeks to roll back "normal" (or routine) form of state intervention associated with the mixed economy and the Keynesian welfare national state (or analogous forms of intervention in the developmental state or socialist plan state) as well as "exceptional" (or crisis-induced) form of intervention aimed at managing, displacing, or deferring crises *in* and/or *of* accumulation regimes and their mode of regulation in Atlantic Fordism, East Asia and elsewhere. It also involves enhanced state intervention to roll forward new forms of governance (including state intervention) that are purportedly more suited to a market-driven (and, more recently, also allegedly knowledge-driven) globalizing economy. This typically involves selective transfer of state capacity upwards, downwards, and sideways, as intervention is rescaled in the hope of securing conditions for a smoothly operating world market and to promote supply-side competitiveness on various scales above and below the national level. ... Finally, as a project to reorganize civil society, neoliberalism is linked to a wider range of political subjects than is typical of orthodox liberalism. It also tends to promote "community" (or a plurality of self-organizing communities) as a flanking, compensatory mechanism for the inadequacies of market mechanism.

Cities play a central role in making neoliberalism operational with regards to the same three main aspects. First, cities hold a strategic role in the context of the market economy because, through the concentration of infrastructure, urban economies are clearly considered to be the engines of economic growth, productivity and efficiency, not only in the national context but also, and especially, at the international level; they are the best place to develop the neoliberal market, based on scale economies and comparative advantages (Burgess *et al.*, 1997; UN-HABITAT, DFID, DPU, 2001; Osmont, 2002; Theodore *et al.*, 2011; Stren, 2012). Cities are also regarded, according to the neoliberal project, as the drivers of greater innovation and competitiveness (Jessop, 2002) and as the place where "public expenditure reduction and economic liberalisation deploy their main impact" (Balbo, 2002a: 18).

Second, from the political point of view, the neoliberal world market can work more effectively if economic and social policies and their supporting institutions are harmonised (Jessop, 2002). In this perspective, cities hold a key role in the reorganisation of the institutional infrastructure, conducive to greater private investments and, in turn, to the smooth functioning of the free market. Cities are the privileged place where the re-structuring of institutions can take place through the "deregulation and reform of regulatory regimes in land, housing, finance, infrastructure, services and employment markets" (Burgess *et al.*, 1997: 20) – such as the elimination of subsidies for, and the privatisation of, public services. As Peck *et al.* (2013) highlight, this institutional remaking on an urban scale is not an end-point in the neoliberal construction but is instead constantly evolving according to external economic, geopolitical and institutional changes and crises through, for instance, the experimentation of "flanking mechanisms" to protect powerful economic actors from market, state and governance failures (Theodore *et al.*, 2011). Within this framework, urban governments are thus called upon to re-organise themselves to take on the strategic task of coordinating this institutional restructuring, which attempts to distribute responsibilities and resources between the needs and the interests of the global level (international competitiveness, liberalisation, privatisation, deregulation, etc.) and the local and regional levels (local economy and well-being, equitable access to services, social inclusion, etc.) (Jessop, 2002).

Third, cities represent a place for civic initiative where, in theory, it is possible to solve the social conflicts and tensions generated by the development of a market economy, through the construction of a neoliberal "market society"[2]; an organisation of everyday life based on neoliberal principles. The involvement of the community sector is thus seen as a contribution to urban efficiency and social inclusion by ensuring "a better grasp of local demand and a better organisation of provided services" (Maloutas and Malouta, 2004: 456), by relying on the new solutions that formal and informal social networks can find in mobilising and organising their own resources, without burdening the scarce resources of the state (Jessop, 2002) but also, most importantly, by "equipping population for autonomy within a modality of government regulated by the incentives and sanctions of the market" (Parker and Debuyne, 2012: 156). As Mayer and Künkel (2012: 208) note, concession by and adaptation of the political and institutional spheres are made to certain groups of people "whose qualities may make them, in a particular

[2] "Market society" is a term utilised by Jessop (2002), drawing on Polanyi (1944, *The Great Transformation*. New York, Rinehart), to explain how liberalism has tried to solve its contradictions and tensions: "… in response to crisis-tendencies in laissez-faire capitalism, many social forces struggle to re-embed and re-regulate the market. The eventual compromise solution was a *market economy* embedded in and sustained by a *market society* (Polanyi, 1944). The same point applies to neo-liberal capitalism. Thus, after the efforts of 'roll-back neo-liberalism' to free the neo-liberal market economy from its various corporatist and statist impediments, attempts are now being made to secure its medium-term viability to embedding it in a neo-liberal market society".

context and however temporarily, useful for generating new markets and new profit opportunities".

Urban governance within the neoliberal project

Along with the crucial role played by cities in securing the construction of a neo-liberal project, urban governance is proposed and taken on as a tool in carrying out this design. This idea rests on three main observations. First of all, the multiplicity of stakeholders characterising the concept of governance (ranging from international enterprises and donor agencies to informal social networks and self-organising communities) are organised through a new hierarchical institutional structure no longer belonging to the welfare state but to the principles of economic efficiency and the managerial effectiveness of the corporate governance. In other words, the old state institutions and hierarchies are progressively replaced with the neoliberal economic hierarchies, under the banner of (*good*, *urban* or *local*) governance (Brown, 2001; Osmont, 2002; Miegeville *et al.*, 2003; Guarneros-Meza and Geddes, 2010; Mayer and Künkel, 2012). Thus, urban governance provides the background to place cities outside the boundaries of the nation-state and make them enter into the global urban network of international competitiveness, where donors, international organisations and private companies play a leading role. Urban governance also aims at organising, at the urban and local level, all of the actors in the public, private and community sectors to integrate their efforts in using the sparse resources available efficiently and, in turn, enhancing urban growth and productivity. Market forces can in effect provide the tools to solve and harmonise social contradictions and antagonisms, "provided they are steered in the good direction through good governance" (Jessop, 2002: 467).

Secondly, following the above assumption, there is widespread agreement that the notion of urban governance has been used to make the neoliberal economic development model objective and inevitable, naturalising and legitimising the market economy on the urban scale. In fact, despite this term bringing with itself a host of issues belonging to the political, technical and social fields, clear priority has been given to neoliberal economic development, while the other factors should adapt to it (Jessop, 2002). According to Maloutas and Malouta (2004: 453), "competitiveness has become the benchmark" for the social problems of cohesion, inclusion, participation, access to goods and services, "making a reversal between goals and means": the improvement of urban life conditions is not considered as an objective per se any more, but a tool to enhance urban economic growth and productivity. Thus, starting from the idea that the neoliberal project does not need justification because it is a natural phenomenon (Brown, 2001; Jessop, 2002; Parker and Debruyne,

2012), urban governance is called in to tackle the issue of social justice within the "horizons" (Jessop, 2002: 456) or "margins" (Maloutas and Malouta, 2004: 454) created by neoliberalism, that is, without impeding competitiveness.

Finally, in the perspective of a neoliberal project regulated by the tool of urban governance, local communities – which represent the local actor involved in the free and competitive market – are called upon to adapt to this natural and spontaneous design. Communities have to rely on their own initiatives and resources to fulfil their needs by developing "family, neighbourhood, informal, or market-based and market-sustaining solutions" (Jessop, 2002: 465). Consequently, urban governance provides the arena in which to manage the proliferation of actors and their innovative solutions through a progressive democratisation of political processes, by applying the instruments of networks, partnerships, participative democracy, pluralistic political system of government, etc. (Devas, 1999; Lombard, 2013). Actually, the lack of resources available and codified rules – where the latter characterises a more open and flexible governance system in contrast, for instance, with the set of laws of representative democracy – permits only the better networked and organised groups to participate in local governance procedures through "a quasi-market process with meritocratic rules", excluding both the powerless voices within social groups and individuals outside the groups (Maloutas and Malouta, 2004: 457; see also Batley and Mcloughlin, 2010; Lombard, 2013). At the same time, Guarneros-Meza and Geddes (2010: 122) underline how the potential emancipatory involvement of community-based organisations can be undermined by the pressure exerted by local governance procedures on such "social movements to adapt themselves to the multiplicity of partnerships and state-sponsored forms of neoliberal citizen participation".

All this represents what is often described as the paradox of neo-liberal governance, where the neo-liberal project purportedly wears a social dimension. Inclusion of civil society actors is defined as "symbolic" (Miraftab, 2004), while the egalitarianism amongst them is considered "simulated" (Jessop, 2002). The purpose is to "re-embed neo-liberalism in society, to make it more acceptable socially and politically" allowing, at the same time, "as few concessions as possible to the forces that oppose program, protagonists, and driving forces of neo-liberalism" (Jessop, 2002: 467). Brown (2001) outlines, for instance, how the democratic aspect of participative governance can indeed embody the private interests of organisations rather than the specific voices of individuals:

> Dans tout ce processus [démocratie], le peuple, compris comme ensemble des citoyens, est le grand absent et on définit déjà l'Ordre Nouveau comme un "gouvernement des organisations, par les organisations et pour les

organisations"[3]. Le grand paradoxe de la gouvernance est qu'on nous propose d'élargir la démocratie à la société civile alors que celle-ci est précisément cet ensemble de relations dans lesquelles les individus ne sont pas des citoyens, mais des simples vecteurs d'intérêts particuliers. ... Ainsi, l'apparente "dévolution" des pouvoir qu'effectue la gouvernance néo-libérale en faveur de la société civile équivaut à exproprier le peuple de sa souveraineté.

In the same vein, speaking about the formulation of community-based waste collection strategies by the municipal government in Cape Town, Miraftab (2004) refers "the paradox of neoliberal governance" to the process of the material exclusion of township residents from adequate paid labour, despite their "symbolic incorporation in development practices and decision making". At the basis of this paradox lies the detection that:

> ...the neoliberal emphasis on community participation, communities' social capital, civil society and decentralization are all closely related ways in which this new governance mode tries to "manage" or offer a "social fix" to the inherent tensions of the capital accumulation *within* capitalistic, racialized, and patriarchal societies (p. 253, emphasis added).

But this is not the only paradox, as also state interests – and not just market and private ones – are winners when neoliberalism adopts techniques of democratic governance. Lombard (2013: 137), for instance, identifies how citizen participation gives rise to two paradoxes within neoliberal democratic urban governance: "first, the power of the state is not necessarily diminished despite the emerging plurality of governance actors; second, the spread of participatory practices does not necessarily lead to citizen empowerment".

1.2. Urban Governance and Service Delivery

Neoliberalism, structural adjustment and the city

In the 1980s and 1990s, the cities in developing countries became the theatre of "a profound shift as neoliberalism has become the dominant development paradigm" (Burgess *et al.*, 1997: 3). In particular, during the 1990s, the international development agenda was further modified according to the neoliberal principles outlined above. This can be recognised in the attempt to embed the economy in a market society – taking into new consideration the issues of poverty and social inclusion in economic development – and addressing the institutional capacities of

[3] Quoted by Brown (2001): Andersen, Svein and Tom Burns, "The European Union and the Erosion of Parliamentary Democracy. A Study of Post-Parliamentary Governance in The European Union", *How democratic is it?*, edited by Svein S. Andersen and Kjell A. Eliasse, SAGE, London, 1996, p. 229.

these countries and not only the functioning of their markets. These reforms have effectively taken place under external pressure from multilateral or international financial institutions – such as the International Monetary Fund (IMF) and the World Bank – which have always had great influence given their role as "financial sponsors of reform" (Batley, 2004). More than just giving expert appraisal and consultancy (Osmont, 1995), it is for instance recognised that the influence of the World Bank did not end at the project level but affected the politics regulating each individual sector (Zanetta, 2001).

Thus, the main instrument of change that external agencies have started to use in these countries has been represented by the macroeconomic development strategies of the Structural Adjustment Programmes[4] (SAP) promoted by the World Bank and the IMF. These programmes were focused also on the cities, in order to improve the productivity of urban economies, strengthen their different and comparative advantages, and turn them into one of the central gears of the free global market (Burgess *et al.*, 1997; Osmont, 2002). Nonetheless, the negative consequences of structural adjustment – notably the widening of the gap between rich and poor – led to further modification of the development agenda during the 1990s, now described as the Post-Washington Consensus (see Öniş and Şenses, 2005).

First, the government's removal of subsidies from services and the cuts in public administration expenditure led to underemployment and a reduction in wages, especially in urban areas (Batley and Larbi, 2004). The resulting serious negative social effects and a general increase in levels of poverty led to the definition of a "second wave" of SAP, characterised by a strong social concern (Osmont, 1999). Thus, beginning in the late 1980s, urban poverty reduction policies began to be systematically included as compensatory mechanisms for the social problems of Structural Adjustment. This took place relying mainly on the emergent definition of a normative framework for development policies and on the progressive construction of a qualitative and participative paradigm for poverty (Ramirez, 2002). Nevertheless, these policies were designed without challenging the discourse on urban productivity and efficiency: improvement in the productivity levels of the poor continued to be seen as the key to urban poverty reduction (Burgess *et al.*, 1997) and, as Burgess *et al.* (1997: 27) highlighted, "it is important to reiterate that the urban poverty alleviation policies must remain just that – policies

4 The structural adjustment mainly entailed the privatisation of state assets, drastic cuts in the number of public sector jobs, the removal of subsidies from basic goods such as food, energy, transport and shelter, the introduction of cost-recoverable prices and of new taxes, and the compression of government social expenditure.

that aim to lessen the number of the poor or to lighten the burden of poverty, no matter what the underlying trends are. They are not to be confused with policies of poverty eradication". In the same vein, Weber (2004: 198) argues that the Poverty Reduction Strategies Papers – which from 1999 represented the lending conditions of the IMF and the World Bank – and the form of governance that they imply are conducive to a "capitalist restructuring", ensuring "the continuity with the SAPs process and the policy (political) framework this entails".

Second, the failings of the SAP as a holistic tool to promote (economic) development led to a re-consideration of the strategic role played by governments, instead of continuing to focus on the widespread recommendations for a reduced presence of the state. Indeed, it became clear that it was no longer possible to just apply for a "smaller" state; on the contrary, it was necessary to introduce a "bundle of management techniques" into the public sector. These have been mainly borrowed from the private sector, and transferred to the public administration under the name of "New Public Management" (Nickson and Franceys, 2003). Batley and Larbi (2004: 40) describe the new approaches in public management, which have involved many developing countries since the 1990s, as follows:

> In the quest for efficiency and effectiveness in government, the adoption of private sector management techniques and practices was advocated to deal with problems of the old public administration by replacing hierarchies with markets in the hope this would minimize the dysfunctions of traditional public administration. In emphasizing management and markets instead of administration and hierarchy, efficiency is considered more important than adherence to rules, effectiveness and results take place over legality, flexibility and adaptation are more vital than predictability and responsibility, and competition with private sector delivery is preferred to delivery by government monopoly and bureaucracy. The new public management (NPM) is a set of reforms meant to shift the public sector from bureaucratic to post-bureaucratic arrangements (Kernaghan *et al.*, 2000)[5].

It is clear how the changes introduced at the government level have essentially been "about liberalising government itself" (Batley and Larbi, 2004), where the market rules sustaining productivity and efficiency have been applied to the institutional sphere. In actual fact, this corresponds to what Jessop (2002) identified as the neoliberal attempt to harmonise economic policies with their supporting institutions.

---

5 Quoted by Batley and Larbi (2004): Kernaghan, K., B. Marson and S. Borins, 2000, *The New Public Organization*, Toronto: Institute of Public Administration of Canada.

Structural adjustment and the delivery of urban services

Urban services have been playing a twofold strategic role in the implementation of Structural Adjustment programmes. Improving access for the poor to urban services means improving the populations' living conditions – health conditions in particular – which constitutes one of the key elements in poverty reduction policies and, hence, of the Structural Adjustment social concern. Secondly, but by no means less important, the improvement of urban services bears a strong impact on urban efficiency. On the one hand, increased access to basic urban services can be carried out through the involvement of the small-scale and informal sector, which will "increase employment for low-income workers as service providers while directly supporting the productivity of users", which the World Bank essentially identifies in the concept of *livability* (World Bank, 2000a: 47). On the other hand, improved urban services represent an essential precondition to making cities more productive and efficient and, ultimately, more competitive in attracting investments (Balbo, 2002b).

Moreover, the changes in public management, upheld by structural adjustment, have deeply modified the organisation and management of urban public services. Public responsibility for social services (such as health, education, water and sanitation) – direct public ownership or state management – was established through reforms that took place from the 1930s to 1950s in Europe and Latin America and in the 1950s and 1960s in South Asia and Africa (Batley and Larbi, 2004). The supply of urban services has been thus decided on the basis of a policy characterised by the model of "public service" for a long time. This approach was based on some relevant considerations: first of all a technical one, as the costs of the construction and extension of infrastructure could be funded only through public finances; secondly a social one, because by subsidising public services it was possible to realise equitable access to essential services and resources; finally an economic one, since efficient urban services entail a well-functioning national economy (World Bank, 1994, 2000a). The public sector's inability to maintain the infrastructure network and to manage it efficiently – for instance, with a well-organised structure of prices or without the system of patronage opportunities for politicians and bureaucrats – produced shortages in service production and provision (Rakodi, 2002).

According to Batley (2004), state responsibility for social services (and economic development) has been challenged by two main waves of public management reforms reflecting, at the same time, the general shift in the international reform agenda. Between the 1980s and the early 1990s, the "first generation" of more broad and general reforms focused mainly on macro-economic stabilisation and on the reduction of state

Box 1.1. Key principles of neo-liberal water policy

a) Water resources should be allocated through the market; that is, private water rights should be created replacing any existing forms of collective or public rights and they should be freely tradable;
b) Water services have to be considered an economic good, in the sense of being a private good that has to be bought in the market; by definition, once WSS are considered to be private goods, non payers can be excluded from accessing them; the notion that WSS are a public or social good must be abandoned;
c) Water services should be provided by private operators, which are inherently more efficient than public ones; if possible, water services should be self regulated by market mechanisms and state intervention should be minimized if not altogether cancelled;
d) Water services are not a natural monopoly, as claimed by the defenders of state intervention; most operations can actually be opened to competition, perhaps with the exception of some core activities; however, high transaction costs can make competition difficult; in these cases, a privately-owned water monopoly is preferable to a public one; even then, keep regulation to a minimum or cancel it altogether if possible;
e) Water users should be transformed into consumers, and right holders into customers.

Source: Castro (2006), chapter 6.

intervention in the management of the economy; the only direct effects on social sector services were the "rises in tariffs and introduction of user fees". These reforms took place under the pressure of a deep economic crisis involving developing countries and the resulting reduction of public expenditure. As previously mentioned, structural adjustment's failure to improve national economies and the resulting reconsideration of the strategic role played by governments, led to a "second generation" of reform, mainly implemented during the 1990s. They focused much more on the organisation of social service delivery and are identifiable "in the organisation and roles of government bodies, in changing attitude toward, and the relationship with, the private sector", involving multiple actors and requiring a long-term commitment from administrative officials and political leaders (*ibid.*: 40). In particular, the public management reforms involving service delivery included: "privatising and contracting out the management of public services, decentralising management to semi-autonomous units within the public sector, the application of charges to users of services, and the development of enabling and regulatory roles by public agencies" (*ibid.*: 34). Batley underlines that, though first and second generation reforms took place separately or simultaneously according to the different developing countries[6], it has always taken

6 For instance, in the four countries analysed in their research, Batley and Larbi (2004) established that in Ghana and Sri Lanka the reform of the public sector followed

longer to implement the "ready-made models of neoliberal economic reform" in the social sectors. Social reforms have been less clear and readily applicable to these sectors than to the economic sectors, and they have faced, in turn, strong resistance from administrative officials and professional staff.

Analysing these changes, Osmont (1999) underlined how the reforms adopted in the urban social services sector, between the late 1980s and the 1990s, presented two main and different features: the technical and financial one, and the institutional one. So, first of all, the reduction in social expenditure particularly affected basic social services and infrastructure by worsening the condition of public services and forcing the public administration to look for alternative solutions to finance them. Above all, the possibility of unbundling the "production" from the "provision" of urban services led to the unbundling of the urban services' delivery process, allowing different providers to get involved in the different phases of this process. This has been more specifically translated into greater involvement from the private sector and increased financial contribution, especially in the phase of production[7] – the water sector representing a particularly intense field of experimentation (Castro, 2006; Marin, 2009; Bakker, 2010; see also box 1.1.). Several forms of market-oriented approaches to the delivery of urban services have hence been adopted in developing countries, such as "increasing focus on the customers, user fees or charges, contracting out of service delivery to the private or voluntary sectors, public-private partnerships, and outright privatisation" (Batley and Larbi, 2004: 49).

According to Rakodi (2002: 144-145), three main arguments have been used to sustain private sector participation. The first relates to efficiency: it argues that the private sector introduces tariffs reflecting the real costs rather than the subsidised tariffs, and that these tariffs can be lower since there is an absence of clientelistic pressure. The second deals with the possibility of improving the fiscal balance of the public sector "by reducing subsidies to loss-making (water) utilities; by increasing tax revenue through expanding the role of the private sector; by reducing the public debt through the outright sale of state asset; or by reducing the government borrowing requirement through requiring the private sector to finance capital expenditure" (Nickson, 1997: 168). The last argument concerns the equity principle: the revenues coming from tariffs reflecting

reform in the economic sector, while in India and Zimbabwe the first generation reforms occurred later and in parallel with the public services reform.

7 See the following literature for more discussion on private sector involvement in service delivery: Harpham and Boateng, 1997; Balbo, 2002b; Batley and Larbi, 2004; Batley, 2006; Batley and Mcloughlin, 2010; Stren, 2012.

the real costs of production and management can, in fact, fund service expansion to more marginalised areas and groups. It is nevertheless clear that the central reasons sustaining the privatisation of urban services have relied on the search for the efficiency of essential urban services. In reality, the possibility for an improved and more equitable distribution of these services amongst the excluded groups is subordinated to the successful collection of fees, which leads to the well-functioning and productivity of service provisions. As stressed below, the failure of the market-oriented approaches has inevitably led to failure in achieving a more equitable access to urban services, once more initiating a search for alternative solutions.

Moreover and most essentially, an institutional re-structuring has supported the technical and financial reforms of urban services. The general context was represented by the new public management, where governments were no longer called upon to physically produce and deliver services ("direct provision"), but simply to ensure that services were available ("indirect provision"), managing, regulating or enabling other providers (Batley and Larbi, 2004). At the city level, this general trend was processed through the widespread application of the subsidiarity principle and, in turn, by management decentralisation. This corresponds with the move from "urban administration" to "urban management", where the latter was supposed to employ alternative methods for, and to organise the new providers involved in, the process of urban service delivery. "Urban management" is thus proposed in order to ensure a more efficient and responsive delivery of urban services, a more direct response to the demands of local citizens and a streamlining of administrative operations. It is based on *efficiency* and *enablement*, rather than on *control*, where the latter characterised the concept of "urban administration" (Stren, 2000, 2012).

Most critical literature agrees on the evaluation that both private actors' participation and market management of public services have produced limited and partial results in improving the quality and efficiency of services and, at the same time, have shown some major limitations (Balbo, 2002a; Budds and McGranahan, 2003; Degnbol-Martinussen and Engberg-Pedersen, 2003, Batley and Larbi, 2004; Barthelemy *et al.*, 2006; McGrahanan *et al.*, 2008). Further to some fundamental and frequently mentioned market failure arguments (see box 1.2.), many critics point to a certain number of weaknesses in the implementation of these reforms. Just to cite some recurrent claims: first of all, the privatisation of state enterprises or contracting out of services has raised costs, and thus tariffs, that deepened the difficulties in accessing services for low-income residents, worsening the spatial and social fragmentation within the city (Balbo, 2002b); in turn, the difficulties in collecting fees have limited the possibility for an improved and more equitable distribution of these services amongst low-income groups, reinforcing their exclusion from the city being "served". Then, decentralised management requires a credible control framework to monitor performance and ensure financial accountability, which is particularly difficult and time-consuming to be built in developing nations; performance management is challenged by the problems of poor organisation of information, lack of enforcement for rewards and sanction as well as weak government commitment to contracts; civil servants' performance evaluations are only partially useful as top officers have little control over the management of services, given the centralised organisation (Batley and Larbi, 2004). In addition, private sector participation has often just changed patronage and corruption frameworks – through the involvement of multinationals – without removing them (McGranahan *et al.*, 2008).

Box 1.2. Market failure arguments

Public goods: Some goods and services are described as by their nature 'public' in the sense that private firms on their own would simply not provide them adequately or at all. These are goods for which there is no means of charging consumers (because non-payers are not 'excludable') and whose consumption does not reduce their availability to others (there is no 'rivalry'). This would occur, for example, in the case of police services, street-lighting or street-sweeping where the benefits are collective. It is impossible to measure how much an individual has consumed and the service in not used up by consumption.

Externalities: The water society would benefit from an extension of a service even though the direct consumers may be unwilling or unable to pay for it: an example of such 'positive externalities' would be vaccination; Producers can pass on costs to the wider society, allowing them to keep their costs to the direct consumers artificially low: an example of these 'negative externalities' would be industrial pollution.

Monopoly: The nature of a good or service or the scale of the initial investment required to prohibit competition – e.g. water supply systems.

Scale and risk: The necessary investments are so large scale or the returns so risky that the private firms are not prepared to undertake them – e.g. major underground infrastructure.

Imperfect information: Consumers and producers have too little knowledge to make informed choices: this sort of situation may arise in the case of professional services (such as health advice) where clients may not be able reasonably to assess the value of options.

Equity or 'merit' good: There are in practice basic commitments in any society to the view that everybody should have access to certain goods and services, regardless of their ability or willingness to pay the market price. 'Government needs to act to provide merit goods because individuals are not necessarily the best judges of what is in their own or the public interest' (Walsh, 1995, p. 10). Education and health typically come in this category.

(Taken from Batley and Larbi, 2004: 24)

In conclusion, urban services made a substantial contribution to promoting the neo-liberal economic model of development, modifying the technical and financial system of their delivery from the perspective of a greater involvement of the private sector, and adjusting the urban and national institutions in sustaining this involvement through the reforms underpinned by the new public management. Even if Batley and Larbi (2004) stress that in developing countries there have been many difficulties in implementing the principal characteristics of the new public management – because of the strong institutional power of bureaucracy – they argue that the social sector reforms have nevertheless been promoted according to:

> the reduction of governments' direct role in managing economies and providing services; greater reliance on markets, communities and individuals to manage services; the adoption by governments of new roles of "steering" (setting policy frameworks, regulating and supporting) service providers rather than providing directly; and the reform of public management to create incentives for efficiency and effectiveness (Batley, 2004: 32).

In this perspective, through the market and political enablement[8] (Burgess *et al.*, 1997), the reallocation of regulatory and productive functions, technical and financial resources, and government responsibilities and tasks was much more conducive to higher levels of economic growth, efficiency, productivity and, in turn, competition – coherent with structural adjustment requirements, according to a neoliberal design – rather than to local social and economic necessities or internal political requirements, driven by the principle of equitable access to urban services.

Urban service delivery and community involvement

Amongst the main changes in the international development agenda of the 1990s, the new interest in community participation in the delivery

---

8 According to Burgess *et al.* (1997), *market enablement* is described as follows: "The state's role in production, ownership, finance, marketing and regulation should be 'rolled back' and its activities restricted to those of 'market enablement'. Government was to be a coordinating and facilitating rather than an interventionist force. Enablement meant facilitating and promoting the formal and informal business sectors and entrepreneurs to provide market solutions for the production, distribution and exchange of urban goods and services" (p. 141). On the other hand, *political enablement* is described as "a transformation in the structure and functions of central and local government, the relations between them and their relations with the market and the community. Political enablement is achieved through political/administrative decentralization, democratization, managerial and institutional reform, the widespread use of NGOs and community-based organizations, and through adopting enabling strategies towards the market and the community in the allocation of material and financial public goods and services" (p. 144).

of urban services certainly represents a major shift. This appeal for community involvement in the development processes was not a new issue *per se*, but the specificity characterising this new attention lies in the relevance of the communities' contributions from a neoliberal perspective of development. In that period, the principle of community participation was revived, in the wake of a new interest in civil society, according to the new development concept of urban governance.

According to Burgess *et al.*, (1997: 152), starting from the 1970s and 1980s, community participation was first acknowledged in *site and services* and *slum upgrading* projects: it was "limited to self-help labour inputs to housing construction, infrastructure installation and some services (e.g. drain cleaning)". In the 1980s, communities began to get involved in the consultation processes concerning the design and plan of urban goods and services. This involvement also rested on the proliferation of the associational life through "tightly-organised systems of social solidarity" sharing common interests and requests, and on the capacity to demand the right to basic services and to protest against environmental dangers (Halfani *et al.*, 1995: 102; Stren, 2000, 2012). Yet, this participation was subordinated to a range of technical choices made in the early planning stages (such as the location of the project, its goals or which community to involve), mainly carried out on the basis of "political allegiances" and subjective criteria of choice, meaning communities were therefore only involved in the final phases of the projects. The resulting increase in community conflicts, inequalities amongst the poor and non-replicability of projects, led to a reconsideration of the role of community participation. Thus, from the early 1990s, communities started to be called upon to directly manage all the processes of service delivery – from planning to maintenance – "according to the decisions and priorities established by themselves". At the same time, this shift had to be sustained by a series of enabling measures and reforms from the central and local governments, aimed at empowering the communities[9].

In this period, the idea of community participation in the delivery of urban services essentially came from the necessity to cope with market failures and the inefficiency of state market led strategies (Miraftab, 2004). As a result, the need has become apparent, not only, to rethink governments' capacities and institutional conditions for sustaining these market-oriented changes, but more importantly to involve communities

9 According to Burgess *et al.* (1997: 151), *community enablement* can be defined as "a strategy adopted by central and local government to coordinate and facilitate the efforts of community and neighbourhood-based organizations to initiate, plan and implement their own projects according to the principles of self-determination, self-organization and self-management".

in the design and realisation of alternative solutions. This is actually one of the key features at the basis of the renewed interest in civil society, from the perspective of a neoliberal state.

> Neoliberal policymakers have reassessed the role of community participation in line with an ideology that stresses the primacy of community and families responsibilities over institutional interventions and the need for the mobilization of resources of a revitalized civil society to fill the resource gaps between need and market and state capabilities (Burgess *et al.*, 1997: 151).

This short paragraph shows how the role that the community is called upon to take in the delivery of urban services is consistent with the "second generation" reforms adopted in the field of social services, as outlined above. As with the private sector, local communities and civil society organisations are now requested to improve the performance of service delivery in two main areas. First, with a direct contribution through their human, technical and financial resources, mobilising cash or in-kind contributions, communities are deemed to "enhance project effectiveness by increasing local commitment" (World Bank, 1994: 78). Second, with their direct involvement in the process of decision-making, which means at the institutional level of urban management, they have to organise themselves in order to reach consensus and enter in the decision-making process, "especially to set priorities for expenditures and to ensure an equitable and agreed sharing of the benefits and the costs of service provision" (*ibid.*: 9).

These two different tasks, belonging primarily to market forces and then to the community sector, are the two sides of the "governance equation" that Harpham and Boateng (1997: 66) identify referring to urban services. They are respectively:

- the *performance* side, concerning the "technical question of how resources are put together to deliver services" and, in turn, the contribution that the community sector can make. This also entails the type of strategies and capacities applied in delivering urban services, which had mainly been decided according to economic debate (notably including those related to the "public sector management, legal framework, economic liberalism and competence to form policies and delivery services"; *ibid.*: 74); and,
- the *representational* side, concerning the "representation of individuals and groups in civil society". This aspect is connected with the relevance that the community sector holds at the institutional level, dealing mainly with the issue of equity within an enhanced participation. It also implies the broad concept of "transparency, accountability, social justice, human rights and democracy" (*ibid.*: 74).

The concept of "urban governance" undergoes, in this way, further changes in the search for solutions to providing urban services, following the approaches based on "urban administration" or "urban management".

Anyway, the debate on the role of community involvement within the urban governance framework tends to be polarised. On one side, there are those who point out how the notion of governance deserves merit for having potentially provided a useful framework that allows community actors to successfully bring about structural improvements in urban development and the delivery of urban services[10]. In particular, Harpham and Boateng (1997: 5) highlight how the notion of urban governance has given emphasis to the "issues of equity and community participation", distinguishing this term from the management techniques borrowed from the private sector as a means of improving the delivery of services and ensuring the opportunity for communities to contribute to service delivery. They also argue that urban governance "should to a certain extent present alternatives to this dominated mode [the capitalistic economy] of what various societies want and can have" (*ibid.*).

On the other side, much critical literature shares the idea that the "governance equation" has simply widened the factors included in the process of urban service delivery – applying it to the principle of an improved equity – without modifying the rationale at its basis. Put another way, the community sector or civil society organisations have just been integrated with or substituted to the direct intervention of the private sector in the operation of urban services, according to a neoliberal state that is no longer responsible for their delivery (Brown, 2001). This view is predominantly connected with the critical analysis of the rationale that some external agencies refer to in funding improved urban services and, in general, in development processes. As an example, the World Bank is supposed to support community (civil society) participation – through the concept of governance – in order to sustain a neoliberal type of development (Osmont, 2002; Lewis, 2003). Thus, for instance, in her investigation of the relationships between governance reforms and democracy in Argentina, Tuozzo (2004: 109-110) argues that:

> The World Bank works with a notion of civil society that is more akin to the Anglo-Saxon experience than that prevailing in Latin America. This particular conception prioritises individual interests over public good or collective rights. In this scheme, civil society is comprised of different individuals binding together with shared interests, often to avoid encroachment by the state. ...civil society is conceptualised in a more instrumental way where society is counterpart of and assume the state's previous responsibilities.

10 See for instance the arguments suggested by Harpham and Boateng, 1997; Hichey and Mohan, 2004; Gaventa, 2004; Joshi and Moore, 2004; Allen *et al.*, 2006a, 2006b; Gaventa and Barret, 2010; Lombard, 2013.

In the same way Miraftab (2004), analysing the community-based waste collection strategies in Cape Town, shows how the neoliberal governance promoted by an international financial establishment such as the World Bank utilises community participation in development processes as a "tool of trade": on the one hand, through the de-politicisation of this concept and the resulting privatisation of the public interest and, on the other hand, through the political concern for social inclusion and the resulting reduction in government responsibility. The loss of the political dimension of community participation, in favour of the technical one, is a recurrent argument in the analysis of how citizen participation is often instrumentalised (see also Hichey and Mohan, 2004; Hichey, 2010). For example, Lewis (2003: 14) highlights that aid donors (such as USAID and the World Bank) support the building of "only an apolitical sanitized version of civil society which excludes organizations and individuals engaged in struggles for political power and influence and which may be critical of neoliberal orthodoxy". In the same way, Parker and Debruyne (2012: 156) underline how "[m]uch like the market, neoliberals imagine community as a self-regulating and thus prepolitical phenomenon. The community could thus, it seemed, provide a field within which to reconcile activist demands for participation with investor and donor demands for privatisation".

Eventually, there seems to be an agreement amongst critics that, although the concept of urban governance can provide room for manoeuvre to allow communities to contribute to structural improvements in service delivery, new and alternative forms of urban/local governance have to be investigated in order to guarantee an equitable provision of urban services to poor and marginalised citizens and to involve the community through a "legitimate and genuinely transformative approach to development" (Hichey and Mohan, 2004: 3), avoiding instrumentalisation and de-politisation and promoting, on the contrary, citizenship.

Informal systems of access to urban water services

Let's now focus on urban water services. In the search for alternative forms of urban/local governance, the options represented by informal systems of access to urban water have been widely investigated in the last two decades[11]. One of the main reasons is that, as Collignon and Venzina

11 The literature on informal water service provisions is really extended; we just mention here some of the most relevant references across the last 15 years: Amis *et al.*, 2001; Jaglin, 2001, 2005; Nickson and Franceys, 2003; Batley and Moran, 2004; Allen *et al.*, 2004, 2006a, 2006b; Mitlin, 2008; Phumpiu and Gustafsson, 2008; Marin, 2009; Batley and Mcloughlin, 2010; Moretto, 2007, 2010; Wild *et al.*, 2012.

(2000) calculated, informal and/or small-scale private water providers serve much of African, Asian and Latin American cities; whilst in some African cities they serve up to 70-80 per cent of the urban population.

Before moving on, a clarifying remark is in order. The word *informal*, in fact, is widely used in literature but there is not a clear consensus as to what it really means (Devas, 1999). On the one hand, as Devas (1999) highlights, the distinction between *formal* and *informal* is difficult to outline because, in practice, "there is an extensive range from the fully formal to the totally informal". On the other hand, in terms of urban water supplies, *informal* is related to both the technical systems of provision (which may include the infrastructure and the sale of water) and the resulting relationships and processes between the actors involved even if the boundaries between these two sides of urban water services are difficult to outline. As table 1.1. shows, many definitions have been given of informal providers, practices, activities, arrangements, networks, focusing from time to time on the different phases, aspects and relationships in the water service delivery process.

Furthermore, it is useful to stress that in many cases the informal systems and arrangements of access to urban water services which belong to the community sector are also grouped under some terms belonging to the wide range of relationships present in civil society. To mention a few examples: *social networks*, *community organisations* and *grassroots organizations*, *livelihood strategies*, *associational life*. According to Amis *et al.* (2001: 19), for example, grassroots organisations are composed by formal as well as informal associations where, in the case of the latter, "informal networks rapidly spread information about what opportunities are available" and "informal insurances ... pay fines and the shared childcare". In the same vein, the kinship networks – comprised in the broad spectrum of civil society associations – in the low-income settlements of Manila are considered "the starting point for many informal networks" aimed at gaining access to water services (Mitlin, 1999: 14)[12]. Similarly, in the references listed in table 1.1., the terms listed above – mainly *social networks* and *community organisations* – are often reused in explaining the social and institutional context and background of the informal practices in urban services delivery. Table 1.2. provides a useful overview of formal and informal practices in water supply.

12 Mitlin (1999) draws on Erhard Berner, 1997, *Defending a Place in the City: Localities and Struggle for Urban Land in Metropolitan Manila*, Ateneo de Manila University Press, Manila.

Table 1.1. Informal water provisions

informal **providers**	'are defined here as illicit or semi-licit – not formally recognised or authorised by government … Most independent actors in the urban water and sanitation sectors are informal, in the sense of not being formal companies, not being authorised or recognised by government and not paying taxes' (Batley and Moran; 2004:5) '… small independent providers … from household vendors of water, small network providers, and private entrepreneurs to cooperative. In some cases they are primary suppliers, and in others they supplement the formal provider' (World Bank, 2003: 171)
informal **suppliers**	'tanker and cart carriers intermediaries' (Nickson and Franceys, 2003: 105)
informal **system/practices**	'water as a gift; clandestine connections; public provisions distorted by bribery; informal sector vendors; water sold from privately owned wells' (WSS wheel in Allen, 2004: 6)
informal **activities**	'are those which do not have full, official recognition or do not comply in some way or other with official procedures and rules … such as drug traffic and illegal electricity connections' (Devas, 1999)
informal **arrangements**	'arrangements that are not necessarily backed up by the formal system but operate on the basis of solidarity, reciprocity or need, such of those cases in which water is provided as a 'gift', by certain members of the community to others in need, but also exemplified in the case of water push cart vendors, who might access water through different means and then sell it to other members of their own community' (Allen *et al.*, 2004: 6) 'informal institutional arrangements … understood here as rule-enforcing mechanisms, including customs, norms and values, religious benefits and social and solidarity networks' (Amis *et al.*, 2001: 14)
informal **networks**	'… such informal organizational networks [the author refers to the social resources of the urban poor: family, kinship, social networks and community-level organizations] have no social, legal or political status.' (Beall, 2001: 364)

Source: Moretto, 2007: 350.

In the face of the increasing contribution of informal practices to accessing urban water services, Wakefield (2004: 2) highlights how informal networks and providers, and the resulting informal arrangements that they produce, "are often relatively underrepresented in donors' policies". This is particularly surprising for donor agencies as, while adhering to the Millennium Development Goals, they committed themselves also to the specific target (number 10) to halve, by 2015, the proportion of people without sustainable access to safe drinking water. Reducing the number of people without sustainable access to water has, in addition, increasingly become a problem related to urban areas, given the growing urbanisation phenomenon that continues to interest developing countries (UN-HABITAT, 2012).

Table 1.2. Overview of formal and informal water supply practices

Provider	Formal practices	Informal practices
Public (state) sector	- Piped network (Household connections and public standpipes) - Wells and bore-wells - Provision by tankers - Water kiosks - Negotiation with communities through 'technical water boards'	- Public provision distorted by bribery practices
Private sector	- Buying from licensed tankers - Buying packaged water (cans, bottles, sachets)	- Buying from tankers - Private vendors drawing from own site piped connections/own boreholes or wells sold directly by bucket or through push carts and bicycle vendors
Community	- Own individual wells and bore-wells - Piped network (community organisation agreement with local authority or public water company) - Piped network kiosks and taps run by the community with NGO support - Boreholes and kiosks run by the community - Horizontal condominiums	- Rainwater harvesting - Water theft - Gifts or paid provision from neighbours - Clandestine connections

Source: Allen *et al.*, 2004: 16.

Previous research on donors' policies for informal water provisions (Moretto, 2007) had highlighted that, despite the significance of the water sector in the international development agenda, policy and strategy documents, together with specifically funded programmes and projects related to informal water supply systems, are very limited amongst international aid organisations[13]. This lack of engagement rests upon three main reasons. The first is due to the fact that these kinds of water service providers have for a long time not been considered as a possible positive contribution to improving urban water access. Up until the end

[13] See also Batley and Moran, 2004; McGranahan and Satterthwaite, 2006; Kjellén and McGranaham, 2006.

of the 1990s they have been in effect mainly opposed or, in the luckiest cases, ignored (Mitlin, 2002). As a direct consequence, they have very seldom been included into the normative framework of development assistance. The second reason concerns the official development assistance structure that makes it difficult to sustain these kinds of providers. Supporting small scale and in particular, informal projects, would require in effect reforming the official development assistance execution of many multilateral aid agencies that "were set up to provide capital assistance to recipient governments and technical cooperation" (Satterthwaite, 2005: 26), and not to support small or community-driven initiatives. Finally, as Batley and Mcloughlin (2010: 148) have stressed, the informal, mutual and often local-level engagements including CBOs and informal practices are hardly scalable-up:

> Donors have left largely untouched the spheres of informal and more local-level dialogue; informal, more mutual agreements between governments and NSPs; and almost the whole sphere of regulation, including the less impositional, more incentive-based approaches described above. Informal, mutual and local-level engagements provide opportunities for learning and the development of trust between state and non-state actors, but, on the other hand, present problems of 'scaling-up'.

Anyhow, although their limited representation in the multilateral aid organisations' policies, on the ground, informal providers continue to allow the poor to gain access to water. At the same time, informal practices and activities generate a range of more or less formal and informal relationships through their interaction with the more formal arrangements and institutions that belong to the public, private or community sectors alike. They are nothing more than one of the many "other forms of collective decision-making" that Harpham and Boateng (1997: 68) identify as characterising the wider concept of urban governance, beyond the government itself (see also Pierre, 2000: introduction). The articulation of these formal and informal processes are deemed to lead to a more complex and horizontal form of governance and engagement, aimed towards create synergies between the different actors and sectors to improve access to water services, rather than scaling down the outputs. Research on improved urban service delivery is thus increasingly focusing on, first, the ways in which informal and community-based practices can articulate to formal systems to provide improved and widely available urban water services and, second, the new and alternative forms of governance that the reciprocal engagement between governments and communities can assume in service provisions (Allen *et al.*, 2006a; Phumpiu and Gustafsson, 2008; Olivier de Sardan, 2009; Batley and Mcloughlin, 2010; Booth, 2011; Gaventa and Barrett, 2010; Wild *et al.*, 2012, for some significant example).

Institutionalised co-production of water services as a way forward

An interesting alternative to conventional approaches in urban service delivery seems offered by the idea of "institutionalised co-production", as a further development of the urban service co-production model in the sense of an articulation of community practices to government systems for service delivery. According to Ostrom (1996), the concept of co-production was initially developed in the late 1970s, as an alternative response to policy recommendations promoting strong centralisation for public service delivery and to the necessity of reducing state expenditures. By co-production, Ostrom (1996: 1073) means "the process through which inputs used to produce a good or service are contributed by individuals who are not 'in' the same organisations. ... Coproduction implies that citizens can play an active role in producing public goods and services of consequence to them". In the same wake,

> Coproduction consists of citizen involvement or participation (rather than bureaucratic responsiveness) in the delivery of urban services. These outcomes are intended to have a positive (rather than negative) impact on service delivery patterns. Coproduction stems from voluntary cooperation on the part of citizens (rather than compliance with laws or city ordinances) and involves actives behaviours. Both groups and individuals may engage in coproduction, but the more important participants from both practical and equity standpoints are collectivises (Brudney and England, 1983: 63).

The roles that different stakeholders can play while contributing to service delivery are clarified by Parks *et al.* (1981, 1999), cited in Vershuere *et al.* (2012: 1085): co-production is "the mix of activities that both public service agents and citizens contribute to the provision of public services. The former are involved as professionals, or 'regular producers', while 'citizen production' is based on voluntary efforts by individuals and groups to enhance the quality and/or the quantity of the services they use". Co-production key assumptions fit thus comfortably in the performance side of the "governance equation" identified by Harpham and Boateng (1997), where citizens are considered a fundamental stakeholder who can mobilise resources for service provisions.

Literature on the co-production of public services has been experiencing an important revival in the last few years and scholars are increasingly investigating the many forms co-production can assume and the numerous facets it can display. Co-production can be the initiative of citizens or governments (Jakobsen, 2012); it may include third sector, public and for-profit organisations (Verchuere *et al.*, 2012); it can be disaggregated into many different services activities such as co-planning, co-design, co-managing, co-delivery and so on (Bovaird and Leoffler, 2012);

it can focus on the operational level – *consumer co-production* – on the strategic one – *participative co-production* – or integrate both of these levels in the form of *enhanced co-production* for public service reform and innovation (Osborne and Strokosh, 2013), just to cite some of the most recent research. In the cities of the South, co-production of basic services is increasingly addressed as a possible way to secure sustainable access to essential services, relying on substantial contributions from both inhabitants and service agents, whether public or private[14]. Echoing Brudney and England (1983: 63) who pointed out as "both groups and individuals may engage in coproduction, but the more important participants from both practical and equity standpoints are collectivities", co-production for service delivery in southern cities generally includes inhabitants not as individuals but mainly through their participation in grassroots, community-based organisations.

With the concept of "institutionalised co-production", a specific form of co-production is identified by restraining communities' collaboration in service delivery to only with state agencies and not with the diversity of professionalised service providers (Joshi and Moore, 2004). Joshi and Moore's definition of co-production is thus focused on "the provision of public services (broadly defined, to include regulation) through regular, long-term relationships between state agencies and organised groups of citizens, where both make substantial resource contribution" (*idem*: 40). Interesting in this definition is the move from a market-centred approach to service delivery towards a state- and community-centred form of development.

Co-production between state agencies and communities – as described by Joshi and Moore – challenges the traditional organisation of service delivery in some important respects. First, "institutionalised co-production" excludes temporary co-production arrangements but foresees long-term arrangements developed on a regular basis. The idea is not here to provide a one-stop solution to a service shortcoming, but rather to build up a long-term process and an enduring engagement between state agencies and communities that can eventually originate positive responses to other weaknesses in service delivery.

Second, the relationship between state agencies and citizens can be "undefined, informal and negotiated almost continuously" (*idem*: 40). The usefulness to move away from standardised contractual and/or

---

14 Just to cite some significant examples of how service co-production in the Global South is addressed in literature: Ostrom, 1996; Batley, 2006; Bovaird, 2007; Mitlin, 2008; Peters and Muraleedharan, 2008; Allen, 2010, 2013; Batley and Mcloughlin, 2010; Moretto, 2010; Booth, 2011; McDonald and Ruiters, 2012; Spronk *et al.*, 2012; Albrechts, 2013; McGranahan, 2013; McMillan *et al.*, 2014.

semi-contractual agreements seems to raise a certain consensus amongst scholars addressing service co-production. This view matches for instance with the "more limited" definition of co-production provided by Batley (2006: 249) where informality is the characteristic of agreements between formal organisations and communities. Informal agreements allow local and informal providers to continue delivering services to the poor without being excluded by tight contractual regulation and, at the same time, without being subordinated to formal agencies (*ibid.*). Additionally, this perspective on non-conventional co-production relationships is in line with Bovaird's (2007: 858) call for a reconceptualization of service provision "as a process of social construction in which actors in self-organising systems negotiate rules, norms, and institutional frameworks rather than taking the rules of the games as given".

Third, power, authority and control of resources is shared between the state and groups of citizens in a way that can entail interdependent and ambiguous relationships as well as blurred boundaries between the public and the private spheres. This means engaging also on the political dimension of resources' management by challenging not only the technical organisation of service delivery, but also its power distribution (McGranahan, 2013; McMillan *et al.*, 2014), contrary to the conventional research on co-production that has focused "on joint forms of service delivery within contextual and institutional analysis" (Mitlin, 2008: 345). "Institutionalised co-production" concentrates thus on new forms of democratic governance and shared decision-making in which power, authority and control are redistributed between government agencies and citizens. It can clearly represent a tool not only to effectively improve service access but also to build up an active and responsible form of citizenship.

CHAPTER 2

Assessing Urban Governance

A Principle Perspective

2.1. Urban Governance Assessments

As previously introduced, urban governance for improved service provisions concerns the sphere of the (new) public management of resources, of the state architecture to organise urban service provision and of the contractual and/or semi-contractual relationships amongst the stakeholders entering into the urban delivery process. But it also concerns the sphere of the unofficial and unorthodox arrangements that are increasingly being developed at the local level to respond to state and market failures in service delivery. Two different but complementary assessment approaches seem thus necessary to investigate the two urban governance facets. One approach should focus on the institutional, organisational and political aspects of urban governance, mainly grounded on a more or less complex dataset of quantitative indicators and/or indexes; the other approach should concentrate on how governance for improved services works in practice through informal, local, and community-based arrangements that are better described through qualitative assessment techniques. Does an urban governance assessment tool able to grasp these two dimensions exist? Let's proceed in a systematic way, overviewing existing governance assessments with respect to their capacity to: first, addressing cities and, then, considering also the informal and unorthodox side of governance arrangements.

If we look at the huge production of governance indicators, we can easily see that measuring and assessing governance encounters many difficulties and an important lack of consensus. A quick look into the UNDP "Global Programme on Democratic Governance Assessments" shows that there is no agreement on a single tool or instrument to assess governance. First, "[t]here exists a plethora of governance indicators, which are used by governments, development agencies, non-governmental organizations, media, academic institutions and the private sector" (UNDP, 2007: 1) to inform users on public programmes, to conduct academic research or to monitor and evaluate governance programmes and projects. For instance, "Governance Indicators: A User's Guide" (UNDP, 2007) and "A User's Guide to Measuring Local Governance"

(Wilde *et al.*, 2009) presents something like 56 different tools for assessing governance. Second, governance assessments also evolved from a "first generation" of evaluation tools focused on programming indicators and academic datasets to a "third generation" emphasising process (Wilde, 2011).

These tools are extremely varied. Part of the assessments concentrates at the country level, are mainly based on statistics and objective measures that include, from time to time, international comparisons of state performance: we can find for instance the "Country Policy and Institutional Assessment" and "Governance Matters V" both developed by the World Bank or the "World Governance Assessment" by the Overseas Development Institute. But we can also find tools focused on some specific aspects of governance, such as the "Bribe Payers Index" or the "Corruption Perception Index" both produced by Transparency International, or the "Press Freedom Index" proposed by Reporters without Borders. On the contrary, another part of these tools is applied to the local level of governance relationships: they usually assess one of the four broad focus areas of governance in a local context represented by local governance (such as the "Local Governance Barometer" by Impact Alliance), decentralisation processes (the "*Desde lo local*" in Mexico, for instance), local democracy (for instance the International IDEA "Local Democracy Assessment Guide") and local government (like the "MIDAMOS" assessment; see Wilde *et al.*, 2009 for many other examples). They usually mix quantitative and qualitative data, often including non-representative sampling and subjective measures such as perception-based evidence collected through surveys and focus groups.

Concerning geographical coverage, there exist governance assessments that can be applied worldwide or in some specific regions and countries: in the first category, we encounter the "Human Rights Indicators" developed by the Danish Centre for Human Rights or the "Local Integrity Initiative" by Global Integrity; in the second category, which is much more ample, there are tools like the "Good Governance for Local Development – GOFORGOLD Index" to be used in Afghanistan, the "Afrobarometer" for Africa, the UNDP "Methodological Guidelines for Local Governance", developed for Latin America or an UNDP assessment, "The Urban Governance Initiative", to be applied in Asia. However, very few of the assessments addressing a more local setting, rather than the country level, explicitly take into account the urban dimension of governance: basically, they do not make any particular distinctions amongst the urban, rural or peri-urban contexts, while proposing a measurement tool for a generic "local government".

Thus, despite the massive production of governance assessment tools, instruments, indicators and indexes, a measure of urban governance, broadly applicable and also open to local governance arrangements seems uncommon. Stewart (2006) suggests that a major reason for the lack of effort directed towards addressing good governance at the urban level rests perhaps on the difficulties related to the data collection process. At the same time, he points out two ambitious attempts made at the international level to cope with this shortcoming: the World Bank database concerning globalisation, city governance and city performance (Kaufmann *et al.*, 2004) and the UN-HABITAT Urban Governance Index (UN-HABITAT, 2004a). It is interesting to note that these efforts are undertaken by two leading international aid organisations, coherent with the dominant role that the donor community has occupied in the governance field since the 1990s (Hydén and Samuel, 2011).

The World Bank database aims at exploring whether globalisation matters for sound urban governance and, in turn, whether globalisation and sound urban governance positively affect city performance. The investigation concludes that governance and globalisation are strictly related and have a positive impact on city performance; urban performance is considered as an output of national economic performance within globalisation while "improving governance at the city level allows cities to translate global opportunity into local value for their citizens" (Kaufmann *et al.*, 2004: 38). This analysis has been developed covering 412 cities worldwide and relying on 35 variables and indicators belonging to other existing databases[1]. To measure city governance, Kaufmann *et al.* (2004: 25) focus on "whether a city is providing services to citizens (low health access gap) and whether it has a well-functioning public sector with low bribery in taxes or in the provision of utilities". In particular, the selection of the city governance determinants focuses on those that can identify the positive link between urban governance and the performance of global cities. This reflects the precise objective of the Bank's database, which is to "investigate empirically what determines the staying power of cities of their performance on a global scale, and whether governance has anything to do it" (Kaufmann *et al.*, 2004: 4).

The Urban Governance Index (UGI) has been developed "in order to enable cities to objectively measure the quality of local governance" (Narang, 2005: 1) through a core set of 18 indicators or 25 short-listed indicators (UN-HABITAT, 2004c; table 2.2.). According to UN-HABITAT, the objectives of the UGI are two-fold: at a global level it aims "to demonstrate the importance of good urban governance in achieving broad development objectives", while at the local level "the index is expected to catalyse local action to improve the quality

[1] The data sources of the World Bank database rely on the UN observatory, EOS database, Taylor database, KLM database. For further details, see Kaufmann *et al.*, 2004.

of urban governance by developing indicators that respond directly to their unique contexts and needs" (UN-HABITAT, 2004b: 11). The UN-HABITAT UGI – belonging to the UN-HABITAT Global Campaign on Urban Governance activities (UN-HABITAT, 2002) – has the overall purpose of contributing to the eradication of poverty, by strengthening and improving urban governance.

Contrary to the World Bank database, selection of the UGI indicators is carried out according to a number of criteria that seem to open the door to an assessment of urban governance where informal and unorthodox local governance arrangements are also taken into account, regardless of their relationships with globalisation processes at the city level as per the World Bank database. This assumption is based on the argument that the UGI – as one of the "flagship products" of the UN-HABITAT Global Campaign on Urban Governance – has to contribute to the overall objective of poverty reduction by relying on the UN-HABITAT definition of urban governance:

> Urban governance is the sum of the many ways individuals and institutions, public and private, plan and manage the common affairs of the city. It is a continuing process through which conflicting or diverse interests may be accommodated and cooperative action can be taken. It includes formal institutions as well as informal arrangements and the social capital of citizens.
>
> Urban governance is inextricably linked to the welfare of the citizenry. Good urban governance must enable women and men to access the benefits of urban citizenship. Good urban governance, based on the principle of urban citizenship, affirms that no man, woman or child can be denied access to the necessities of urban life, including adequate shelter, security of tenure, safe water, sanitation, a clean environment, health, education and nutrition, employment and public safety and mobility. Through good urban governance, citizens are provided with the platform which will allow them to use their talents to the full to improve their social and economic conditions (UN-HABITAT, 2002: 14).

These paragraphs highlight two features of urban governance particularly close to the way in which governance works on the ground, for instance through institutionalised service co-production. First, the idea that urban governance is built upon the principles of inclusiveness and citizenship is in line with the possibility for citizens to enjoy a share of the power, authority and control over resources, as suggested by the (new) form of citizenship within service co-production. UN-HABITAT (2002: 13) overtly advocates this option when it states "governance is not government. Governance as a concept recognizes that power exists inside and outside the formal authority and institutions of government. In many formulations, governance includes government, the private sector and civil society". Second, the emphasis on the informality of arrangements and relationships fits comfortably in the kind of "undefined, informal and negotiated almost continuously" relationships, suggested by Joshi and Moore, which characterise the way in which co-production operates. Finally, as suggested by Shafiei (2011: 49), "the global implications of this method [*the UGI*] and its origination and introduction by one of the best acknowledged international bodies further justify the selection of this method" to evaluate urban governance.

2.2. Urban Governance Index (UGI) Principles and Indicators

If the idea of urban governance on which the UGI is built upon seems hence to be addressing the two facets of this notion – by including formal, institutional processes and the more informal and locally-based arrangements to access urban services, both contributing to generate citizenships and inclusion – is this Index able to provide a measurement of these two dimensions, in an integrated way? To provide an accurate answer to this question, we need to go into the principles and indicators composing the Urban Governance Index.

UGI principles

The overall UGI organising structure rests on the "Five Core Principles of Good Urban Governance", based on the seven principles characterizing the rationale behind the Global Campaign on Urban Governance, but also on the norms and principles shared by the Agencies within the UN system (UN-HABITAT Global Campaign on Urban Governance, 2001; table 2.1.). These five core principles have been ultimately selected on the basis of their simplicity and clearness in expressing governance key concepts and on the basis of their relevance to the cross-cutting principle of inclusiveness. UN-HABITAT (2004a) stresses how the choice of these principles is linked to Amartya Sen's five measures of freedom (Sen, 1999) whereas each principle may be used to evaluate one of them: *Effectiveness* for Economic Facilities, *Equity* for Social Opportunities, *Participation* for Political Freedom, *Accountability* for Transparency Guarantees and *Security* for Protective Security. As a matter of fact, the relationship between UGI core principles and Sen instrumental freedoms[2] rests on "the more *inclusive* idea of capability deprivation" as a barrier to poverty understanding and development realisation, rather than the "*exclusive* concentration on income poverty" (Sen, 1999: 20, emphasis added).

2 Sen (1999: 36-37) defines the "instrumental freedoms" as a means to "contribute to the expansion of human freedom in general, and thus to promoting development" (*principal means*). They are opposed to the "constructive freedoms", which represent the value of freedom as an end of development for itself (*primary end*).

Table 2.1. UN Agency norms on urban governance

Urban Management Programme UNDP/UN-HABITAT	The Urban Governance Initiative / UNDP	UN-HABITAT
1. efficiency 2. equity 3. participation 4. sustainability	1. accountability 2. consensus 3. equity 4. effectiveness 5. efficiency 6. participation 7. responsiveness 8. rule of law 9. transparency	1. civic engagement and citizenship 2. efficiency 3. equity 4. security 5. subsidiarity 6. sustainability 7. transparency and accountability
UNHCHR	**UNDESA**	**UNICEF**
1. transparency 2. responsibility 3. accountability 4. participative government responsive to the needs to the people	1. accountability 2. fairness 3. efficiency 4. effectiveness 5. non-discriminatory 6. transparency	1. universality 2. indivisibility 3. participation 4. accountability

Source: UN-HABITAT Global Campaign on Urban Governance, *Minutes of Inter-Agency Meeting on the Norms of Good Urban Governance*, New York, June 2001; UNICEF, 1998, *Guidelines for the Human Rights-Based Programming Approach*; UNDESA, *Green procurement in Developing Countries*; UNDP, *The Urban Governance Initiative*.

In the context of the UGI development, UN-HABITAT elaborated these five core principles to work out a measurable definition that could allow and justify the selection of indicators linking policy objectives and governance principles. Eventually, only four of these five principles were recommended to be part of the Index[3].

1. *Effectiveness (including: efficiency, subsidiarity and strategic vision)*

Effectiveness of governance measures the exiting mechanisms and the socio-political environment for institutional efficiency (through subsidiarity and effective predictability) in financial management and planning, delivery of services and response to civil society concerns (UN-HABITAT, 2004a: 21).

In the UN-HABITAT perspective, effectiveness is primarily related to the principle of *institutional efficiency* which links directly to the quality

[3] The Five Core Principles of Good Urban Governance were initially recommended to be adopted in the Urban Governance Index (see UN-HABITAT, 2004a). Nevertheless, after the second Field Test Report, the *security* principle was not recommended to be part of the Index because of its weak representation with regards to the selected evaluation criteria (see box 4.1. and UN-HABITAT, 2004a).

of processes and institutions involved in the delivery of public services and in promoting local economic development on the one hand, and to the sound management of revenues sources by the local government, on the other. Responsibilities for an efficient delivery of public services[4] and the promotion of local economic development have to be delegated at the closest appropriate level, according to the principle of *subsidiarity*. This should bear the benefit of, first, allowing citizens to be more involved in the process of urban governance and, second, to respond more effectively to local needs and priorities, in the hypothesis of a local government with sufficient resources and autonomy to meet its responsibilities. Finally, in strict relation with this last point, the presence of a *strategic vision* elaborated by the local government is considered to be the local administration's first step commitment towards the welfare and priorities of citizens and thus the responsiveness to their needs and desires.

2. *Equity (including: sustainability, gender equality and intergenerational equity)*

> Equity implies inclusiveness with unbiased access (be it for economically weaker sections, women, children or elderly, religious or ethnic minorities or the physically disabled) to basic necessities (nutrition, education, employment and livelihood, health care, shelter, safe drinking water, sanitation and others) of urban life, with institutional priorities focusing on pro-poor policies and an established mechanism for responding to the basic services (UN-HABITAT, 2004a: 22).

According to UN-HABITAT, the equity principle is considered two-fold: equity in access to the basic necessities of urban life and equity in decision making according to the principle of participatory governance. The access to basic services and to decision making has to be open to all categories of citizen – from a *gender equality* perspective to an equal consideration of formal as well as informal economic sectors – according to an inclusive approach to urban management and development: "[t]he sharing of power leads to equity in the access to and use of resources" (UN-HABITAT, 2002: 13). UN-HABITAT also includes the concept of *sustainable development:* a sustainable management of urban areas is considered an essential part of this principle, where the sustainability is seen in a classical perspective as the balance between the social, economic and environmental needs of present and *future generations.*

---

4 Shafiei (2011: 36) explains that "[t]he notion of efficiency compares the quantity of a service provided (e.g. tons of refuse collected) to the resources (e.g. labour hours) used to provide it, and thus efficiency provides a measure of how reasonable service costs are".

3. *Participation (including: citizenship, consensus orientation and civic engagement)*

> Participation in governance implies mechanisms that promote strong local representative democracies through inclusive, free and fair municipal elections. It also includes participatory decision-making processes, where the civic capital, especially of the poor is recognized and there exists consensus orientation and citizenship (UN-HABITAT, 2004a: 23).

The participation principle rests on the idea that good urban governance allows all people to take part in the urban development process as *citizens*, to whom the rights and the conditions under which to effectively contribute to the decision-making process are recognised; it requires, likewise, that citizens be willing to engage themselves in public life as active and not passive actors. As a result, participation includes two core aspects: representative democracy and participative democracy. The former is directly related to competitive and regular elections of mayors and of municipal councils as well as the essential political entitlement associated with voting rights and the participatory selection of legislators and executives. The latter gives a measure of civic engagement through which citizens might not only contribute to the common good, but also reveal their needs and priorities and, in turn, influence decisions that are going to affect them. UN-HABITAT stresses how an effective participatory democracy demands a real responsiveness to civil society concerns.

4. *Accountability (including: transparency, rule of law and responsiveness)*

> Mechanisms are present and effective for transparency in the operational functions of the local government; responsiveness towards the higher level of the local government; local population and civic grievances; standards for professional and personal integrity and rule of law and public policies are applied in transparent and predictable manner (UN-HABITAT, 2004a: 21).

Within the broad issue of who is accountable to whom and for what, UN-HABITAT emphasises the accountability of local government to civil society. According to Sen's instrumental freedom related to transparency guarantees, accountability means that an individual is aware of what the institutions offer and what he can expect to receive, based on conditions of trust, openness and lucidity of interactions inside civil society. In particular, three sub-principles are at the basis of UGI selection. Firstly, the transparency sub-principle relates, in one aspect, to the free availability and direct accessibility of information to people affected by the decisions made concerning public interests. This includes, for instance, that enough and easily understandable information is provided for municipal contracts, tenders, budget and accounts. In another aspect, transparency is linked to an open and organised citizen consultation, through which the local authority shows a willingness to accept and relies on advice and criticism from community members to modify its practices. Secondly, the integrity (or rule of law) sub-principle involves a responsible and reliable staff of public servants, from a professional and personal point of view. An integrity assessment may rest on anti-corruption facilities such as disclosure of public officials' assets and regular independent audits – what is generally defined as "horizontal accountability" (Ackerman, 2004). Finally, the responsiveness sub-principle is assessed with regards to central government and citizens ("vertical accountability"), in order to reduce possibilities for local authorities to dismiss complaints (Rakodi, 2003): "if central government is able to close local authorities or remove councillors at its discretion, councillors are likely to be more accountable upwards rather than downwards to their citizens" (UN-HABITAT, 2003c: 22-23).

UGI indicators

Urban Governance Indicators have been chosen according to the 4 basic principles listed above. The final set of UGI indicators consists of 25 indicators short-listed – including the 18 core set of indicators (see table 2.2.) – chosen against the criteria of *ease of collection*, *universality*, *relevance* and *credibility*. These indicators have been selected through UN expert group meetings and are characterised by the following features (UN-HABITAT, 2004a; Narang, 2005):

- UGI indicators focus on the process of decision-making (e.g. involvement of civil society in a formal participatory planning and budgeting process before undertaking investment in basic services) rather than on inputs, outputs or outcomes of these processes.
- They intend to be credible and robust in order to properly measure the quality of governance and to be sufficiently universal to enable comparison.
- They try to tackle both the issues of universality and contextualisation. According to this purpose, UN-HABITAT proposes as a useful approach identifying "core" indicators relevant across different countries and contexts and "satellite" indicators specifically suited to each country's particular context.
- Indicators should allow stakeholder participation and local ownership in the data collection process and in the following application of these indicators to the decision-making process.
- UGI indicators focus on quantitative data collected at the city level, even if qualitative information is considered important to complement the core set of quantitative indicators.

– In addition, UGI indicators have been selected trying to disaggregate national/sub-national indicators from the local ones. This has been done in order to take into account urban governance issues at the national level but also at the local one that is the closest level to citizens.

Table 2.2. Comparison between "core" and short-listed UGI indicators

Principle	High ranking or "core" indicators	High and select moderate ranking indicators
Effectiveness sub-index	1. Local government revenue per capita 2. Local Government transfers 3. Ration of mandates to actual tax collection 4. Published performance standards	1. Local government revenue per capita 2. Ratio of actual recurrent and capital budget 3. Local Government transfers 4. Ratio of mandates to actual tax collection 5. Predictability of transfers 6. Published performance standards 7. Customer satisfaction survey 8. Vision statement
Equity sub-index	5. Citizens charter 6. Proportion of women councillors 7. Proportion of women in key positions 8. Pro-poor pricing policy	9. Citizens charter 10. Proportion of women councillors 11. Proportion of women in key positions 12. Pro-poor pricing policy 13. Street Vending
Participation sub-index	9. Elected council 10. Mayor selection 11. Voter turnout 12. People's forum 13. Civic Associations (per 10,000)	14. Elected council 15. Locally elected mayor 16. Voter turnout 17. People's forum 18. Civic Associations (per 10,000)
Accountability sub-index	14. Formal publication of contracts, tenders, budget and accounts 15. Control by local government 16. Anti-corruption commission 17. Disclosure of personal income and assets 18. Regular independent audit	19. Formal publication of contracts, tenders, budget and accounts 20. Control by local government 21. Codes of conduct 22. Facility to receive complaints 23. Anti-corruption commission 24. Disclosure of personal income and assets 25. Regular independent audit

Source: UN-HABITAT, 2004a.

Although the purpose of the Urban Governance Index is to measure the quality of governance in the form of the mechanisms, process and institutions through which this concept is put into operation, placing "emphasis on the quality of relationships and processes between key stakeholders at the local level" (UN-HABITAT, 2004a: 18), it seems unfortunately to fail in assessing the more informal, unrecorded and narrow-extended dimension of local governance arrangements. The UGI indicators rely in effect on quantitative and factual data and information to be collected at the city level with the support of municipalities and local governments (UN-HABITAT, 2004a; Narang, 2005). Nonetheless, as we have showed, many current urban governance systems, in place at the local level, are based on informal arrangements between formal institutions and practice-rooted community activities, to both plan and manage common urban affairs and to organise access to basic services, as previously described in the context of "institutionalised co-production". Yet, these informal arrangements – and their resulting outputs – are quite often left as unofficial, unrecorded and dispersed information which, by its nature, is not systematically collected and organised into administrative databases. As a result, a gap can clearly occur between the practice-based, mainly informal, local governance processes to produce improved urban services, and the formal governance prescriptions and mechanisms for service delivery within the institutional sphere. Given the UGI focus on official quantitative data and information, informal local arrangements – despite their theoretical and conceptual recognition – can be excluded from an urban governance assessment.

However, based on the hypothesis that the assumptions underpinning the Urban Governance Index – basically, the principles of inclusiveness and citizenship that embrace both formal institutions as well as informal arrangements – are consistent with those behind local governance arrangements for improved water services, as described within "institutionalised co-production", we assert that the urban governance framework (principles and indicators) offered by the UGI can be a valuable reference tool to also grasp how governance for improved service provisions works at the local level, within and around low-income communities. As a result, a more neighbourhood-level and qualitative approach needs to be developed, in order to integrate and verify the UGI quantitative and city-level assessment. This means considering the multitude of actors present at the very local level, ranging from official service providers – such as state agencies operating at the local level or private enterprises – to local governments, community-based organisations or organised groups of citizens, and all inhabitants.

2.3. Qualitative Applicability of the UGI to Institutionalised Co-production for Water Services

The applicability of the UGI urban governance framework at the community level is tested through the possibility of identifying a set of qualitative queries corresponding to the four UGI core principles and aimed at assessing local governance arrangements characterising "institutionalised co-production". To do so, we conducted a review of the UGI principles that is not intended to be a complete analysis of the many uses, interpretations and implications of these principles, but rather a look at a possible way in which these principles may be made operational at the community level, with a special focus on water service delivery. The UN-HABITAT understanding of urban governance is taken as the guiding standpoint to qualitatively develop and adapt them, while selected literature is used to better develop their operationalization. This is worked out through the identification and clarification of some criteria belonging to each principle, against which the qualitative analysis is built and tested. The four UGI principles are reviewed below while the set of queries emerging from this review are presented in chapter 4, table 4.2.

Before we move on to this review, a definitional and conceptual remark about these four principles is in order. As with the word *governance*, for the concepts of *effectiveness*, *equity* – and especially – *participation* and *accountability*, a widespread agreement about their meaning does not exist. This lack of consensus is primarily due to the fact that they are often used as overlapping, interchangeable or also contradicting concepts (Beazley, 2000; Rakodi, 2003; Myllylä and Kuvaja, 2005; Shafiei, 2011); secondly, they are sometimes transposed into different application fields, bringing yet another definition to their original meaning and value (e.g. the company connotation of *corporate governance* or *financial accountability*); thirdly, they are often linked to an ambiguous determination of the context in which they are applied (for instance, are we referring to participation by community members, by the private sector, or by local associations? Is the accountability of whom to whom for what?). The following pages will try, when possible, to clarify as much as possible how these concepts are used, when they overlap with the others and the specific context in which they are translated.

Effectiveness

Two criteria are proposed by UN-HABITAT to evaluate the effectiveness of an inclusive governance process: the mechanisms in place for the "effective delivery of public services" and for "responsiveness to the society"[5]. The first criterion relates to the effectiveness of water supply and focuses on its delivery side. Effective delivery of urban water services is essential as it can contribute to a reduction in poverty levels by improving the living conditions of the urban population – especially

[5] The measurement of institutional efficiency explicitly concerns the municipal level and is thus left to the quantitative UGI indicators.

their health – and by enhancing urban efficiency through the reduction of the economic and social impacts of bad health conditions (Nunan and Satterthwaite, 2001). In the light of possibilities for the co-production of basic services, at the neighbourhood level, the efficient delivery of water supplies ought not solely to be linked to sound and cost-effective administrations, to which responsibilities are delegated (according to the subsidiarity principle), but should also include informal contributions from the community sectors, with enhanced possibilities to also bring about responsibilities – beside the obvious right to access to services – over service delivery.

The second criterion refers to the "responsiveness to society" or "to civil society concerns". Responsiveness from local authorities does not need to rest exclusively on the idea that a citizen is solely a client or customer, thus resulting in merely an assessment of service delivery performance. The UN-HABITAT standpoint, in fact, seems to adopt the governance metaphor of "two hats for one lady" (Vigoda, 2002) in which the idea of "serving clients" is combined with that of "collaborating with citizens" as partners, through a variety of collaborative and cooperative arrangements that redistribute not only responsibilities, but above all power and resources among institutions, agencies of public administration and citizens, in the form of individuals or groups[6]. The UN-HABITAT's support for a responsible, as well as collaborative approach to urban service delivery is clearly explicit in its approach to good urban governance where "individuals and institutions, public and private, plan and manage the common affairs of the city. It is a continuing process through which conflicting or diverse interests may be accommodated and *cooperative action* can be taken. It includes formal institutions as well as informal arrangements ..." (UN-HABITAT, 2000: 5, emphasis added).

It is interesting to note that both criteria pertain to distributional considerations primarily concerning the sharing of national or local economic resources, but also the sharing of power, resources, dialogue and responsibilities with citizens. In this way, the effectiveness principle – within the UN urban governance framework – rests not only on an

---

6 In Vigoda's metaphor (2002), the lady represents the "public administration" while the two hats are associated with the two different kinds of relationships between the citizens, government and public administration: "responsiveness" or "collaboration". With this metaphor, Vigoda (2002: 530) distinguishes between two competing options: "(1) There are two substantially separated faces of government and public administration (two ladies), one that adopts the idea of responsiveness and one that favours collaboration; (2) the discipline of governance and public administration is more coherent (only one lady) than we might think, and at most it changes colours over time (two hats)".

economic perspective related to revenue sources and service performance, but includes a wider perception of an effective collaboration between local authorities and citizens as partners, in order to respond to their need for urban services and for inclusive methods too. As a result, the qualitative analysis at the neighbourhood level should focus on how governance arrangements developed within co-production have brought changes in the effective delivery of water supply, on the one hand, while improving responsiveness to community water needs and collaboration amongst the main stakeholders in guaranteeing an effective water delivery, on the other. Community members, organised groups of citizens, as well as the other local stakeholders involved in the service delivery may provide evidence as to the results of service co-production and processes in assessing and reacting to community needs.

Equity

Two main criteria also characterise the equity principle: "tools or mechanisms present for (equitable) access to basic services" and "equity in decision making". The choice of these two equity criteria in urban governance processes is consistent with Sen's argument about the relevance of social opportunities[7]: "[t]hese facilities are important not only for the conduct of private lives (such as living an healthy life and avoiding preventable morbidity and premature mortality), but also for more effective participation in economic and political activities" (Sen, 1999: 39). This two-fold perspective on the equity principle reflects the distinction between distributional equity – represented by a wider and more equitable distribution of public service – and procedural equity – concerning the participation of all groups in decision-making (Smith and Hanson, 2003).

Distributional equity focuses here on the presence of mechanisms at the local level able to guarantee equitable water access. The notion of access to basic services in the cities of developing countries is divided by Crane and Daniere (1996) into "community" versus "individual" access and "supply" versus "demand" access. In particular, the supply/demand issue addresses the distinction between "access costs" (reflecting the supply factors) and "access values" (reflecting the effective demand) where these values – the local behaviours towards water service[8] –

[7] According to Sen (1999: 39), "[s]ocial opportunities refer to the arrangements that society makes for education, health care and so on, which influence the individual's substantive freedom to live better".

[8] Amongst these demand behaviours, which are related to waste disposal in the paper's example but allow generalisation to water supply, Daniere and Takahashi (1999: 273) list: from the community perspective, "population participating in environmental community projects … population willing to contribute resources to improving overall

are strictly dependent on the service price structure[9]. Accordingly, mechanisms able to guarantee an equitable access to water supply should address both individual and community access and should plan an inclusive access structure combining service costs and water-related values alike. Within institutionalised co-production, a community's demand and value for water services are at the core of local governance arrangements for improved provisions. This consents in principle to work out a local and flexible payment system, based on communities' previous economic experiences accessing water and on local capacity and willingness to contribute to the service creation and to pay for one type of disbursement over another – rather than on a unique model of cost-recovery for low-income communities.

In regards to the second criterion, procedural equity is directly connected to the idea of participative governance (see below) and, hence, to the inclusion of men and women in the decision making process, but also marginalised and disadvantaged groups, as well as formal and informal actors. On the one hand, procedural equity should contribute to supporting local authorities in comprehending and identifying the processes which entail distributional equity (Smith and Hanson, 2003). This can be achieved, *inter alia*, through reducing the communication gap between authorities and users in order to define appropriate levels and types of basic service, to identify affordable provisions and to improve the two-way responsibility between administrators and citizens. On the other, following the UGI equity indicators, gender equity has to be considered as significant also at the settlement level: the presence of men and women in co-production, for instance, could be unequal or quite well-balanced but with different roles.

On the ground, a governance assessment of service co-production has to investigate the equity principle in relation to both the issues of cost for and value of water supply, and the way local stakeholders are included in the decision-making process with respect to these issues. This might be performed through two main sets of queries amongst the different actors involved in the governance process for water supply:

- community members should be asked their opinion on the improvements they have experienced in water access (quantity, quality, frequency, affordability, etc.) and any possible desirable

environment of community … population claiming that pests and vermin are a major community problem". From the individual perspective, the authors note "willingness to pay for a service (for one type of service over another) or for better services".

9 The water access values affected by the price structure are identified by Crane and Deniere (1996) in the quality and quantity of water consumption, health consequences, time spent to get water, efforts to obtain and transport water amongst others.

improvements, in order to identify and understand the values they assign to their water service and the processes put in place to effect these improvements. It would also be significant to investigate – on a case-by-case base – the role of women, children and marginalised groups in service delivery.

– other actors involved in the delivery process should be asked about the role of communities (community-based organisations as well as individual citizens) in providing water supply to the settlement. These questions aim at exploring how the local governance arrangements in place can arrange a water supply system capable of tackling the demands of the whole community and how they contribute to providing an equitable supply of drinking water. Local stakeholders also have to state their attitude with respect to the cost and method of payment for water, in order to verify the existence of, or to assess the possibility of establishing, a local and flexible payment system, based on the community's values and resources.

Participation

UN-HABITAT (2004a: 22) considers that "the level and quality of participation determines the quality of governance". In the same vein, the relevance of participation within urban governance is clearly revealed by Stewart (2006: 198): "... while indicators of other qualities may be important, measures of citizen participation are essential to any complete assessment of good urban governance, as they provide some idea of the extent to which interests are considered within an urban community and whether this consideration may or may not be equal"[10]. Here the UN-HABITAT Index once more identifies two main criteria: "representative democracy" and "participative democracy", claiming that both of them create participation. However, regardless of this distinction, all the UGI participation indicators focus on the presence and measurement of formal organisations and procedures intended to enable participation, rather than on the assessment of an effective and direct community participation in

10 There is an unquestionable and widespread consensus over the increasing importance of community participation in urban development policies and processes over the past 30 years but also, at the same time, over the unclear definition of this notion. Community participation is often deemed a justification for the government not to assume the cost and responsibility of service provisions and to rely on volunteer efforts and materials (Ackerman, 2004; Swyngedow, 2005). In this sense, chapter 1 has already illustrated how much critical literature shares a common position with regards to non-effective and market-oriented community participation – promoted amongst others by some international aid agencies (see Moretto, 2007) – in the context of urban governance and urban service delivery.

itself. They thus avoid taking into account the differentiation between a "formal level of participation" and "the way that participation operates in practice" (Tunstall, 2001).

For this reason, a focus on the informal aspects of governance arrangements in co-production should primarily address the mechanisms belonging to participative democracy rather than those appropriate to official representative democracy. When investigating governance for the co-production of water services one must then consider, first, how community participation is structured around organised citizens and/or community based organisations and, second, whether and how these organisations are able to have a voice within local governments and service providers (even if this second point is tackled in a broader way within the principle of accountability). Although the presence of a representative democracy is considered a pre-requisite for any participatory mechanism (Devas, 2001; Navarro Yànez, 2004; Brewer-Carías, 2010), the assessment of institutions and procedures to enable participation, within the UGI, cannot be left only to the quantitative indicators.

As a result, the extent of participation in practice – as opposed to participation through formal procedures – can be investigated, assessing community members' level of awareness *vis-à-vis* CBOs activities to co-produce water services and the level of their formal/informal involvement in these CBOs' works, with respect to participation to their internal organisational structure and functioning, and with respect to the community members practical contribution to improved water provisions. With regard to the participation aspect of "representative democracy", questions should be asked of organised groups of citizens in order to understand how and through which procedures they represent the community in front of local authorities and service providers.

Accountability

Unlike the UGI quantitative indicators, a more qualitative governance exploration at the community level should consider two accountability levels – both downward and upward (Oia, 2011): the first concerns how water service providers (whether state agencies, private enterprises or local governments, to which the responsibility for service delivery is often decentralised) are accountable to the community based organisation involved in co-production, while the second focuses on how CBOs involved in service co-production are accountable to the community (and vice versa). It follows from this that the three accountability criteria identified by UN-HABITAT (transparency, integrity and responsiveness) should be considered in respect to both these levels.

The concept of "co-governance for accountability" (Ackerman, 2004) may help in exploring accountability at the local level. Ackerman argues that the notion goes beyond a light form of participation intended to devolve costs and responsibilities for services over to communities; on the contrary, co-governance for accountability "confuses the boundary between state and society. In addition to co-producing specific services and pressuring government from outside, societal actors can also participate directly in the core functions of government itself" (*ibid.*: 451). Assessing the extent to which governance arrangements built around community-based organisations for service co-production have been successful in securing the active and permanent involvement of societal stakeholders is fundamental (see also Benequista and Gaventa, 2011, for the idea of "blurring the boundaries" between local actors in the tiers of governance).

Again, according to Ackerman (2004)[11], "answerability" and "enforcement" are the two main aspects characterising the concept of accountability. The former is related to the information that the decision-makers should communicate to those who will be affected by such decisions (*transparency*). The latter concerns the mechanisms required for guiding the correct decision-makers' behaviour and responsibility for their actions (*integrity and responsiveness*). Transparency has to be investigated at the community level with regard to the information that each community member gathers from the CBOs involved in co-production. Integrity and responsiveness are related to local public servants in a horizontal and vertical way, but at the same time, to the CBOs' capacity to address improved access to water for the whole community; to go beyond their particular interests in obtaining the service for themselves or for a reduced area of the settlement; to work with other community-based organisations for a wider settlement development outside the hierarchies and leaderships within the community; to behave as the spokesperson and the defender of all community needs in front of the local administration and the water providers.

11 Ackerman's way to consider accountability of public officials is drawn on Schedler A., "Conceptualizing accountability", in Schedler A., Diamond L. and Plattner M.F. (eds.), *The self-restraining state: power and accountability in new democracies* (pp. 13-28), Boulder: Lynne Rienner.

Chapter 3

Co-producing Water Services in Venezuela

3.1. Operational and normative framework for water supply in Venezuela

Institutional framework for water provisions

The state-owned water enterprise INOS (*Instituto de Obras Sanitarias*) was established in 1943 to provide water and sanitation to urban areas, as stated in the *Plan Quinquenal 1941-1946* (Marcano, 1993). This national body was characterised by a highly centralised planning and management system designed to supply water to the whole nation and by its prioritisation of sanitation and social solidarity over service quality or economic efficiency (Corrales, 2004). When the INOS was closed in 1991, the strategy was to decentralise service provision to the municipalities that were legally in charge of it, since the 1989 Organic Municipal Government Act. The idea was consistent with the wave of decentralisation programmes endorsed by the government after the *Caracazo* (1989) to tackle the severe deterioration in basic services, which had partially arisen because of the fall in oil revenues during the 1980s (Goldfrank, 2004). Decentralising services to local governments was also part of the general modernisation process set off in the Venezuelan economy in the 1990s to challenge the centralised and shareholder model of service delivery (Corrales, 1998, 2004). Changes in water service should reflect, in effect, the economic and institutional neoliberal adjustment programmes launched in Venezuela in the late 1980s[1], and focus thus on regulatory frameworks to enhance efficiency and quality in service provisions through the introduction of market mechanisms, on financial tariffs and subsidies strategies to make services autonomous and on the promotion of private initiatives[2].

At the beginning of the 1990s, the service was temporarily entrusted to the ten regional water companies – under the holding company

1 Two structural adjustment packages were implemented in Venezuela: the first in 1989 and the second in 1996 (Smilde, 2011).

2 In 1992 there was an attempt to contract out the provision of water and sanitation services in the Capital Region to a private operator, but eventually, no successful proposal by the private sector was made (Coing, 1995).

HIDROVEN – and to the *Corporación Venezolana de Guayana*, until the eventual transfer to municipalities. In reality, municipalities were unwilling to assume service provision responsibilities because of their poor financial conditions and the absence of municipal investment plans together with incentives to charge for the service (IDB, 1998). This situation had not changed when the 2001 Organic Water and Sanitation Law (*Ley Orgánica para la Prestación de los Servicios de Agua Potable y Saneamiento* – LOPSAPS) established that responsibilities for the provision and control of water services should be transferred to municipalities or metropolitan districts by the end of 2006 (LOPSAPS, 2001: art. 11 and 134). According to this new law, local governments could decide to associate themselves with other municipalities (*mancomunidad*) in order to organise water services, if technical, economical or regional solidarity criteria apply (*ibid.*: art. 12). Municipalities have to collaborate with states and the central government as "the municipal, state and central executive powers shall perform their responsibilities for the development of water and sanitation services in a harmonic and coordinated way and in accordance to this law [LOPSAPS], its regulation and the applicable norm" (LOPSAPS, 2001: art. 8).

Municipal governments represent the most important institutional level in water supply as they should provide the service (directly by the municipality or the metropolitan district, by associations of municipalities, or by public, private or mixed enterprises); they have to define the management model, the involvement of private companies, the tariffs, the regulation, control and sanctioning procedures for service provision. The state level can participate in the development of the service in non-urban areas and support municipalities technically and economically (infrastructure as well as subsidies). The national level carries the responsibility of guaranteeing the creation of technically and economically sustainable water management units (*Unidades de Gestion*) and to manage the production of untreated waters. To do so, the law predicted the establishment of three new bodies.

- The National Office for the Development of Drinking Water and Sanitation Services, intended to deal with water provision policies and strategies together with the definition and finance of water investments and the subsidies structure belonging to the Fund of Financial Assistance;
- the National Superintendency of Drinking Water and Sanitation Services, which would be responsible for the control of water quality, water tariffs and maintaining relationships amongst the stakeholders involved together with the design of the contracts' models;
- the National Management Company, which would be in charge of the production and sale of water.

Table 3.1. Municipal, state and national responsibilities according to the LOPSAPS

	National government	**States**	**Municipalities and metropolitan district**
Politics	Fulfilment of national objectives		
Planning	Plans and objectives at the national level		Contribution to the national power
Regulation			
– Sectoral structure	Approval of management units; National Management Company		Investment plans; proposal for management units or associated services (*mancomunidad*)
– Commercial, technical	General norms for service delivery		Specific norms
Vigilance	Officials behaviour		Definition of fines
Control (sanction)			
Financing	Financial politics and financial contributions	Financial contributions	Financial contributions
– Tariffs	General model for tariffs		Approval of tariffs on the basis of the general model
– Subsidies	Design and financing	Financial contribution	Financial contribution
Operation	Production and contracting out of water operations	Support in formal and informal settlements	Direct or indirect provision; management criteria selection
Rural aqueducts		Support in management	Promotion of community management and technical assistance
Technical assistance	Provision	Provision	Promotion of educational programmes
Private sector participation	Participation support		Service delegation; promotion of the water boards
Transfer	Support for concluding the process by the established date		Shaping management units and applying for revision

Source: Corrales (2004: 24).

Table 3.1. presents a schematic summary of the municipal, state and national responsibilities in the delivery of water and sanitation services according to the 2001 Organic Water and Sanitation Law. But, the new regulatory structure based on these three national bodies has not yet been created[3] and HIDROVEN is still the only regulatory body for water and sanitation in Venezuela (IDB, 2011; Higuerey *et al.*, 2012; Martínez, 2013). The deadlines set for creating the aforementioned new management bodies and to transfer responsibility from HIDROVEN to the municipalities have been delayed through successive amendments to the LOPSAPS (LOPSAPS, 2007). Higuerey *et al.* (2012: 14) highlighted, just two years ago, that "up to now, neither the bodies have been created nor have new processes for the transfer of the service to municipalities been opened. For the first one, the deadline is almost reached while for the latter one only two years are left".

Water service provision organisation and distribution

After the closure of the INOS in 1991, operations concerning the extraction, treatment and distribution of water have been shared amongst different bodies within the national territory. Currently, the biggest provider is represented by the state-owned holding HIDROVEN (*Compañía Anónima Hidrológica de Venezuela*), which operates through its nine hydrological water companies, delivering water to around 73% of the Venezuelan population (IDB, 2011). In addition, seven decentralised water enterprises, dependent on state and/or municipal governments, and the *Corporacion Venezolana de Guayana*, provide drinking water in the remaining Venezuelan states and other local areas (Higuerey *et al.*, 2012).

According to the WHO / Unicef Joint Monitoring Program for Water Supply and Sanitation estimations (JMP, 2014), piped drinking water (water "piped onto premises") was available to 85.4% of the Venezuelan population by 2005, which allowed HIDROVEN to state that Venezuela had already achieved the Millennium Development Goal water target of halving the proportion of people without access to safe water by 2015. The 2011 national census revealed further improvements in piped water access (91% of the Venezuelan population, JMP, 2014)[4]. Nonetheless, access to water is largely uneven as it significantly varies between urban and rural areas (respectively 94% and 75% according to the 2001 census, if considering "improved water sources")[5], while the percentage of piped drinking water with respect to other sources is also different in urban and rural areas (84% and 53% respectively; INE, 2011) and between formal and informal settlements (respectively 88.1% and 59.9%, data belonging to the 2001 census; Corrales, 2004).

3 As per December 2013.

4 Besides, Corrales (2004) stressed how, after the tremendous improvements achieved by the INOS originally, progress in the service's coverage was very limited during the 1990s in respect to the increasing trends in the growing population.

5 No official summary statistics for the 2011 Census.

Despite important improvements in the cost recovery of service expenditures, the water supply system in Venezuela seems to show many structural shortcomings. Water rates in the service organisations are still highly unaccounted for (about 60% of the total water production) largely due to water losses in the infrastructure system, illegal connections, free water delivered to informal or difficult to reach settlements and to a widespread culture of non-payment for the service. In addition to this, the percentage of employees in the hydrological companies is very high with respect to the number of service subscribers, and tariffs are exceedingly low – often frozen by central governments (1981 and 2003) – covering only 40% of costs and undermining the financial viability of the service providers. There is, therefore, a large dependence on central investments for the infrastructure's maintenance and expansion (Lacabana, 2003; Corrales, 2004; Escalona *et al.*, 2009; IDB, 2011).

Actors involved in the water sector in Venezuela

Many stakeholders are involved in the water service delivery process. At the public level, the water management and governance framework is composed of the institutional bodies directly involved in the water sector, such as the centralised (Hidroven and its water companies) and decentralised hydrological enterprises, as well as the central government bodies listed above (the National Office for the Development of Drinking Water and Sanitation Services, the National Superintendency of Drinking Water and Sanitation Services and the National Management Company). Other national institutions also operate in the water and sanitation sector – although their activities are not limited to this area – such as: the Ministry of Environment and Natural Resources (MARN), the Ministry of Infrastructure (Minfra) and the Ministry of Health and Social Development (MSDS).

Some international financial organisations (such as the World Bank, the Inter-American Development Bank and the *Corporacion Andina de Fomento* – CAF), big national and international enterprises working in the major public works, together with smaller ones are also involved in the water and sanitation sector. They deal principally with maintenance works, water meter reading, invoicing and payment collection, and distribution by tankers and in bottles. In addition to this, a multitude of micro-enterprises and cooperatives operate in the water sector, in a more informal way, forming a flourishing "social economy" (Corrales, 2004; Lacabana *et al.*, 2004; Cariola and Lacabana, 2005b).

Finally, within the community sector, it is possible to identify a number of different stakeholders who contribute to improving water provisions. All of the citizens – especially residents of low-income settlements – are called upon to enter into the water delivery process, especially in the financing and construction phases. Also, the *Mesas Tecnicas de Agua* and the *Consejos Comunitarios de Agua* represent *ad hoc* organisations encouraging people to look for – and realise – water provision alternatives. In addition, the Local Public Planning Councils and other community based organisations get involved in water and sanitation debates and projects as the bodies in charge of channelling citizen participation in municipal affairs.

3.2. Technical Water Committees as drivers of "institutionalised co-production"

Decentralising water provisions to municipalities is seen as a way to increase efficiency in water delivery but also to promote citizen participation: "[t]he transfer of water and sanitation services to municipalities – as per the LOPSAPS – aims at guaranteeing the service to Venezuelans, allowing them to participate in the construction of alternatives for the autonomous solution of their problems" (Hidroven and MARN, 2004: 8). In order to participate in the search for and realisation of improvements in accessing water, citizens are requested not to act individually but to be organised in locally-based associations that can mobilise most of the local potentialities and knowledge while working for the benefit of the whole community. As Cariola and Lacabana (2005b) have highlighted, since the end of the 1990s, citizen participation in improved water provisions has evolved from a variety of individual practices and traditional forms of community-based associations, to new and organised forms of participation, especially for residents in low-income settlements. In the case of water and sanitation, the *mesa tecnica de agua* (technical water committee) represents a unique and fundamental community body in the service delivery process.

Technical Water Committees (Mesas Tecnicas de Agua – MTA)

The MTA constitutes a community-based organisation, built with the purpose of channelling community participation in the decision-making process and in the physical improvement of service delivery. The *Mesas Tecnicas de Agua* (MTA) – "water boards", "water committees" "water technical tables" or "water technical fora" – are not a product of the current government. The first MTA was established in 1993, in the Antimano *parroquia* within the Libertador Municipality in Caracas. According to Arconada (1996), the municipality and the water company Hidrocapital both acknowledged the importance of community participation in the delivery of drinking water and agreed not only to directly inform citizens about service distribution but also to change procedures and attitudes in order to involve them in the service delivery. Two main results were achieved through what has been coined "Antimano's experience". First of all, residents in the Antimano parish were provided with an aqueduct, serviced by a pump station and a storage tanker. Second, through the provision of the service, local institutions indirectly recognised citizens as dwellers of the metropolitan area and no longer as illegal invaders.

"Antimano's experience" has been a positive one, taken by the state water company operating in the capital district (Hidrocapital) as an example of the successful joint management of water service improvements. Since the arrival of the Chavez government, MTAs have entered into the legal and formal water provision system, considered to be "the expression of the citizen participation's new politics" in the water sector (Lacabana *et al.*, 2004). From 1999 onwards, HIDROVEN established a Citizen Participation Program in order to replicate and strengthen the experience of the water boards at the national level. (Corrales, 2004; Hernaiz, 2004; INDIA, 2006). Community participation in water and sanitation service delivery is now institutionalised in the form of the *Mesas Tecnicas de Agua*, as recognised in the LOPSAPS (2001).

From an operational perspective, MTAs carry out a series of activities within the settlement in order to improve access to water. In particular, they have to set up:

- a map of the settlement including the existing water pipelines (formal and informal). Community participation in this phase is essential as only residents have knowledge of the existence and location of illegal piped connections;
- a water census or survey in which the community collects general data about each household in the settlement and their way of accessing water;
- an inventory of problems (*diagnostico*) and the area's own subsequent water project, which are jointly analysed and designed by the community and the hydrological company, thus benefiting from both local knowledge and technical expertise to realise a sustainable network.

Once these activities have been realised, the construction of the water project involves workers from the hydrological enterprise together with the settlement's residents. Generally, the water company funds and installs the main pipe through which the informal settlement is connected to the formal water network. Afterwards, the settlement is organised into *condominiums*, which are smaller areas sharing one or a few secondary pipelines. Company workers and settlement residents realise these secondary water connections together. Consequently, within the *condominiums*, each household connects its shelter to the piped water network through hand-made plastic connections. *Condominiums* are not just a physical settlement subdivision through which to organise the piped network's construction: they are also a social unit around which the whole process for improving access to water (water survey, network construction, network maintenance, payment collection, etc.) is built. They are the basic settlement units where common problems concerning water and sanitation facilities are discussed and decisions over solutions are taken.

The results of this joint work between institutions and communities are threefold. First, water distribution usually improves according to an equitable criterion as the service can now reach the low-income settlements where water was not previously provided. Although water is not available 24 hours a day in low-income settlements, the planned rationing (*racionamiento programado*) of the service allows residents to know when they can access and, thus, collect water and, in turn, to organise their time without being totally dependent on the resource's availability (Cariola and Lacabana, 2005b).

Second, this new form of community participation reduces the socio-spatial segregation of the poor. Through such a kind of community organisation people move from a sort of "privatisation" of their social life (Lacabana *et al.*, 2008: 11) where problems were solved at the household or settlement level by "an individualistic and competitive approach to service access" (Moretto, 2010: 94), to a new appropriation of the public sphere that takes place not only in the collective physical space of the community, but also at the institutional level, which is now open to community participation (Moretto, 2010; McMillan *et al.*, 2014). This change contributes to the creation of a new form of citizenship in which "the *Mesas Tecnicas de Agua* ... are an instrument for developing responsibilities, knowledge, claims for rights and citizen construction" (Cariola and Lacabana, 2005b: 126).

Third, local governance arrangements develop around MTAs as a result of collaboration and cooperation with the other main stakeholders in the drinking water provision process – notably, the water companies, the municipalities and all of the community members. The recently designed institutional framework for water provisions implies an increased number of actors and organisational bodies involved in water design and management – such as the "Communal Management" (*Gerencia Comunitaria*) – which in turn entails the invention and re-invention of new relationships, agreements, and decision-making processes amongst these actors.

Table 3.2. Distribution of MTAs by Hydrological entreprise and year

Administration	Enterprise	2006	2008	2009	2010
Centralised	HAndes	73	253	253	269
	HCapital	553	700	711	712
	HCaribe	129	355	531	699
	HCentre	220	450	677	774
	HFalcon	328	675	849	696
	HLago	616	1,518	1,686	1,813
	HLlanos	51	186	233	234
	HPaez	34	131	132	274
	HSuroeste	137	404	506	587
Decentralised	CVG-GOSH	15	66	69	72
	AMerida	128	191	220	239
	AMonagas	64	229	282	296
	APortuguesa	85	205	248	248
	AYaracuy	162	241	298	346
	HLara	105	314	402	402
	AEjido	43	64	87	90
	HBolivar	40	261	286	358
	Total	**2,783**	**6,243**	**7,470**	**8,109**

Source: Higuerey *et al.*, 2012: 31.

Although it is not easy to quantify the actual impact of this new system of access to water, as water boards are still growing and developing in Venezuela and official data is not widely available (table 2.2. provides an estimation of the MTAs' evolution and distribution, while the Venezuelan Ministry of Citizen Power for Environment considered that in 2013 the total figure rose to about 9,700)[6], the results seem to be encouraging. For many low-income residents, the quality and the frequency of their access to water has improved, construction and connection costs have been dramatically reduced, and new agreements and relationships have developed around this new community-based organisation (Arconada, 2005; Cariola and Lacabana, 2005a; Lacabana *et al.*, 2008; Lopez Maya, 2008; Moretto, 2010; Lopez Maya and Lander, 2011; Allen, 2012; McMillan *et al.*, 2014).

Technical water committees as alternative governance arrangements for the "institutionalised co-production" of water services

Organisational arrangements between the state-owned water companies (the affiliated hydrological enterprises within the Hidroven

[6] Information obtained in the Ministry of Citizen Power for Environment website (http://www.minamb.gob.ve, accessed on the 25th of March 2014).

holding) and the structured groups of citizens represented by the technical water committees can be considered a form of "institutionalised co-production" for urban water service delivery. First, they are based on relationships between the organised citizens (the water committees) and state institutions (the water companies): they are an expression of the current "strategy to reach social equilibrium" through the inclusion of two fundamental actors (Lopez Maya and Lander, 2011: 64): "first, the state, in all of its administrative levels and branches, as the creator of conditions that make possible the empowerment of citizens; second, citizens, who through their participation in families, communities, and organised groups are transforming themselves into political subjects with values such as solidarity, respect for democracy procedures, and co-responsibility".

Second, long term co-production is ensured in many respects: these relationships run alongside the process of mapping and surveying, designing and realising the project, managing and maintaining the water network, and collecting payments. Plus, MTAs are one of the CBOs playing a leading role in Venezuela's wider socio-political project, based on the principles of social inclusion and participatory democracy, which clearly aims to both continue to exist and to be developed further in the future. The technical water committees not only constitute a "fundamental tool of Municipal Councils ("*Consejos Comunales*") with respect to their experience in the execution of projects and in the management of funds"[7] (Lacabana *et al.*, 2008: 18), but they are also supposed to interact with other like-minded communities to build up Bolivarian socialism. Lacabana *et al.* (2008: 17) clearly indicates how

> the MTAs, as a mean of community participation, cannot be endorsed in itself but ... have to be articulated with the other social community organisations[8] not only to improve quality of life in an integral manner and to contribute to the construction of a new institutional context envisioned as a relation between the community and the State, but also to support the political construction of new forms for the exercise of political power.

7 With respect to the resources available to improve water and sanitation services, it is interesting to notice that the Venezuelan government has established a special fund to directly finance the projects designed by the communities themselves through the MTAs. According to Flores (2006: 26), through this Financial Fund (*Fondo de Financiamiento*), the executive power invests in water projects presented and approved by the MTAs, relying on the principles of self-reliant participation, co-responsibility and co-management. At the end of 2006, the fund had provided more than 200 water community projects with about 42 billion bolivars (about 20 million US dollars).

8 Such as the *comité de tierra urbana* to deal with urban planning and housing; the *comité de salud* to provide settlements with health facilities, the mesa *tecnica de energia* to install a power network.

Third, these relationships are mainly informal – based on personal contacts amongst community members and between community members and public institutions – and they evolve during the implementation of the water project. As we will further expand upon, nuances in such kind of relationships are possible. While it is true that over time, the MTAs have been institutionalised and formally codified as forms of participation in the current *Ley Orgánica para la Prestación de los Servicios de Agua Potable y Saneamiento*, many relationships between the MTA leaders and water company officials continue to be unofficial, informal and often based on personal contacts. The possibility that such informal and evolving relationships could exclude certain communities or social groups from the co-production system does exist. While unofficial, informal and occasional relationships between water officials and community leaders can provide rapid ad hoc solutions to the many different water shortage situations, at the same time, they risk side-lining two other central actors: first, the municipal administration (legally responsible for water supplies), which works through more bureaucratic relationships and, second, the residents themselves, as a significant number are not always easily reachable through informal arrangements. The following chapters will analyse how these dynamics take place on the ground.

Fourth, in their extensive studies concerning the Venezuelan water committees, Cecilia Cariola and Miguel Lacabana have emphasised that the joint work between the state agencies and communities is based on an attitude of co-responsibility towards developing solutions for improved water supplies.

> … there is a substantive progress in the creation of a new concept of organisation and community participation…: this does not concern a matter of looking for an answer to the State neither to substitute its responsibility, but it is a matter of assuming concretely a co-responsibility in the solution of problems … to build a new network of relationships between the State and the community, helping to build social tissue in the popular settlements (Lacabana *et al.*, 2004: 28).
>
> The relationship with the institutions is seen from a perspective of greater responsibilities by communities, where the latter carry not only the right to ask for the service, but also a series of obligations for its procurement and functioning: it stems from that a vision of co-responsibility between the community and the institutions … communities not only ask for and are clients of a service, but they also held in its construction and are a part of it (Cariola and Lacabana, 2005b: 125-126).

This co-responsibility for defining and producing water and sanitation services is also shared by the institutions:

> within the Ministry, we see the water boards as an instrument to promote the participation and the *protagonismo* of organised communities in the service's

> improvements: drinking water delivery, sewerage, treatment and disposal of used water ... development of a shared management in order to establish a new culture of water based on the co-responsibility for the service (Tirado, 2006: 6).

This sharing of responsibilities and resources is evident throughout the whole process: from the outset, organised communities and state water agencies cooperate to develop the water network. They jointly define and decide on the water project, fund and construct the network and organise the payment collection. As stressed by Mitlin (2008: 352) with respect to self-organised co-production, "the objective [of co-production] is not simply to develop a model that is passed over to state employees to manage and/or to take over as the state gains in capacity" to deliver services, but it is also "to build up a new form of citizenship based on the hypothesis of participatory democracy and finalised to a social inclusion process" (Lacabana *et al.*, 2008: 12). This kind of co-responsibility in co-production represents a way of dividing power, authority and control over resources between the state and its citizens.

Water Community Councils, National Water Meetings and Cooperatives

Citizen participation in the water sector is not limited to the search for water shortage solutions at the settlement level, but also at the municipal and even national level. The Water Community Councils (*Consejos Comunitarios de Agua – CCA*) and the National Water Meetings on community experiences on drinking water and sanitation (*Encuentros de experiencias comunitarias en agua potable y saneamiento*) represent alternative formal and institutional opportunities in which MTA representatives are invited to share their experiences with other MTAs, as well as representatives from the elected authorities and from the hydrological enterprises. These meetings aim at sharing best practice with other communities and deepening knowledge on the subject for a more equitable access to water (Lacabana *et al.*, 2004; Cariola and Lacabana, 2005b; Higuerey *et al.*, 2012).

The Water Community Councils group a number of communities accessing water from the same hydrological systems (e.g. basins, reservoirs, pump stations, etc.), so that they can share similar problems concerning – for instance – water supply frequency or quality. Participants include MTA leaders, public servants from local institutions and technicians from the hydrological enterprise working in that area.

> The council has two main features: it is public and it meets regularly. This means that is open to all citizens with no exclusions, and it meets at a regular time and in a place well known. It has three main functions:

a) It prioritises issues from all identified needs. Taking into account available resources, it establishes a working plan according to priorities established collectively.
b) Organises programmes. Both the water company and the community take on commitments. An agenda for the workers is established and resources are allocated to each task.
c) Follow up. The main task of the council is to exert social control over the public company. Meetings always start with the same questions: What have we agreed? What did we do? What didn't we do and why? And so on (Arconada, 2005b: 133-134).

In the Water National Meetings, MTA representatives meet with other water boards and cooperatives, members of the ministries involved in the water supply (such as Minfra) and Hidroven officials. Besides communicating their experiences, MTAs discuss issues concerning the water sector at the national level – e.g. the normative framework for water and sanitation, the role of drinking water with respect to the current socio-economic changes, the role of education, etc. (Hidroven, 2005).

Another form of citizen participation in the water sector is represented by the cooperatives:

> The drinking water service's administration in Venezuela is managed by the Hidroven subsidiary enterprises. Nevertheless, most operations are contracted out to cooperatives. ... In many country's areas, the hydrological enterprises hired cooperatives which are engaged in the water sector. There are more than 200 cooperatives dealing with the water operation, commercialisation as well as maintenance, and more than 4 million people from the cooperatives (engineers, technicians and workers) have benefited (Tirado, 2006: 6).

In many cases, after realising the water network, an MTA organises itself into a cooperative in order to organise payment collection and maintenance[9] for the network. The involvement of MTA members in these operations – through the cooperatives – entails a more efficient water supply management, as they already know the organisation, function and weaknesses of the water network in their settlement. At the same time, cooperatives represent a working opportunity for low-income residents within the framework of a more "social economy" (Lacabana *et al.*, 2004). Moreover, according to Arconada (2005a: 135) "worker co-operatives are growing in the water sector as part of a process of

9 This information was collected through interviews with MTA leaders, the General Director of Environmental Education and Community Participation Department at the MARN and the ex-community coordinator for Hidrocapital of the Capital District of Caracas.

eliminating intermediaries between those who do the work and the company that has to formulate, plan and inspect this work. They are also a sign of the increasing control of organised communities over their basic services".

CHAPTER 4

Applying the Urban Governance Index to Institutionalised Water Service Co-production

4.1. Water service co-production within two low-income settlements

New governance arrangements based on institutionalised co-production for water services will be analysed here, through focusing on two case studies located in the southern periphery of the Metropolitan Region of Caracas, represented by the Tuy Valleys (*los Valles del Tuy Medio*). Maps 4, 5 and 7 show the location of these two communities (*Paso Real 2000* and *Hacienda el Carmen*) within the respective municipalities (Cristobal Rojas and Paz Castillo) in the Tuy Valleys. After briefly presenting the Tuy Valley sub-region and the current technical organisation for water supply, a description of the two analysed communities is provided to present their history, the tenure situation, the composition of the MTAs and their relationships with the local, regional and national bodies. Further detailed information about the community's developments and achievements in improving water provisions is also offered and analysed in chapter 6. A short introduction related to the two municipalities where the communities are located is also presented: additional details about the financial and administrative situation of these two municipalities are extensively covered in chapter 5 and in Annex 3.

The Caracas Metropolitan Region and the Tuy Valleys

Over the last five decades, the metropolitan city of Caracas has experienced a growth in population, activities, functions and relationships which has resulted in its expansion into cheaper land, beyond its topographical barriers (Cariola and Lacabana, 2007). The Caracas Metropolitan Region (CMR) is now composed of the Metropolitan Area of Caracas located in the central valley and an additional four geographical sub-regions: the Tuy Valleys are located in the sub-region to the South (the other three are the *Altos Mirandinos*, the *Litoral Central* to the North and the *Guarena-Guatire*; see map 1). The CMR accounts today for about five million inhabitants (Lizarraga, 2012; Delgado Linero, 2013). From an administrative perspective, the CMR is divided into three entities: the Capital District, the Miranda State and the Vargas State. The Tuy Valleys belong to the Miranda State and are organised into 6 municipalities (map 2).

Map 1. The Caracas Metropolitan Region

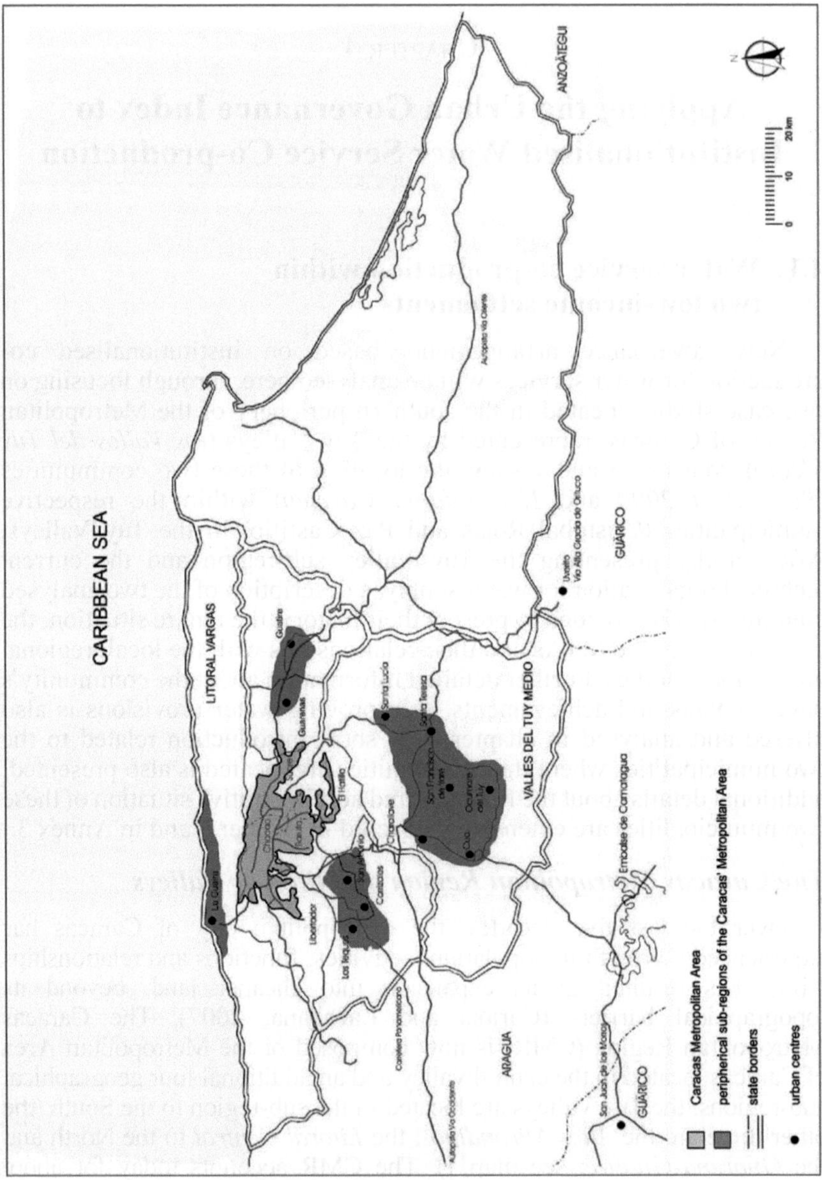

Source: adapted from Cariola and Lacabana (2007).

Map 2. The Municipalities of the Tuy Valleys

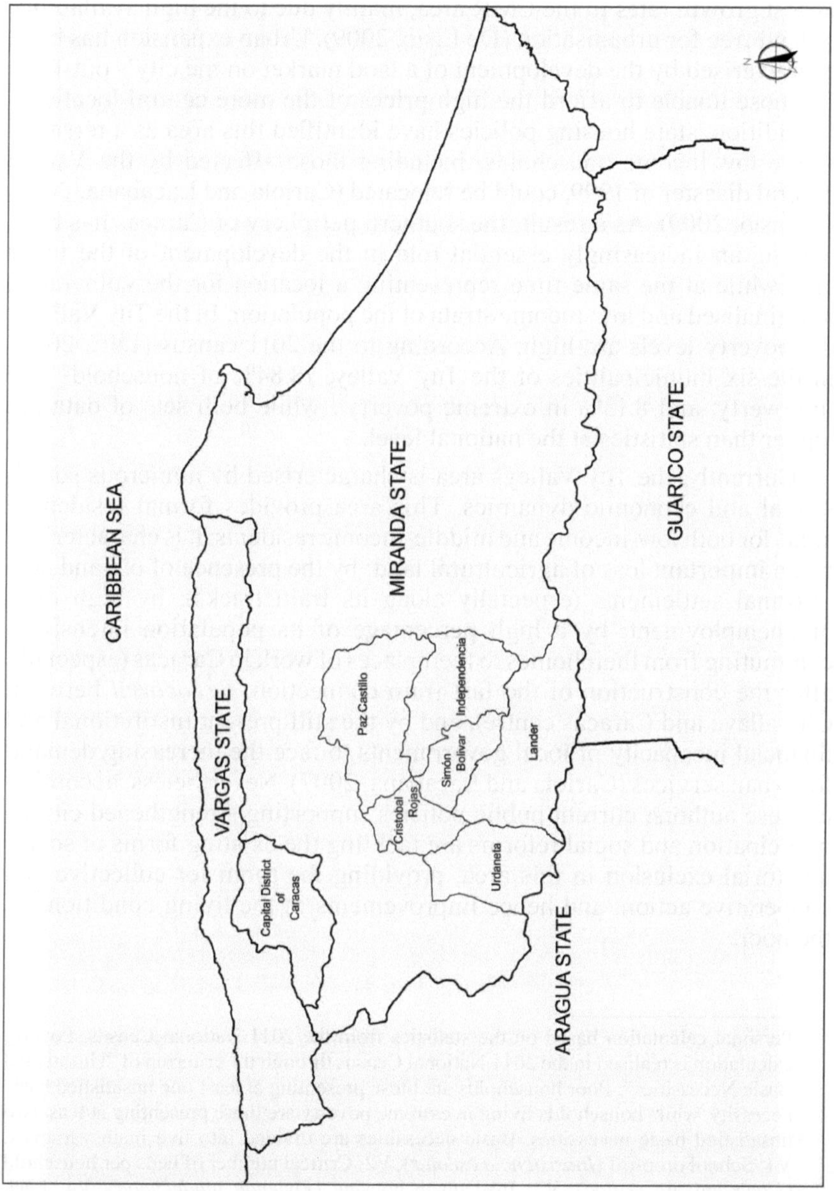

Source: adapted from Cariola and Lacabana (2007).

By the end of the last century, the Tuy Valleys sub-region had one of the fastest growth rates in the CMR area, mainly due to the high availability of land free for urbanisation (De Lisio, 2009). Urban expansion has been characterised by the development of a land market on the city's outskirts for those unable to afford the high prices of the more central locations. In addition, state housing policies have identified this area as a territory where low-income households, including those affected by the Vargas natural disaster of 1999, could be relocated (Cariola and Lacabana, 2007; De Lisio, 2009). As a result, the southern periphery of Caracas has been playing an increasingly essential role in the development of the inner city, while at the same time representing a location for the vulnerable, marginalised and low-income strata of the population. In the Tuy Valleys, the poverty levels are high. According to the 2011 census (INE, 2011), in the six municipalities of the Tuy Valley, 18.84% of households live in poverty, and 8.13% in extreme poverty[1], while both sets of data are higher than statistics at the national level.

Currently, the Tuy Valleys area is characterised by numerous social, spatial and economic dynamics. This area provides formal residential areas for both low-income and middle-income residents; it is characterised by an important loss of agricultural land; by the presence of old and new informal settlements (especially along its train tracks); by high rates of unemployment; by a high percentage of its population intensively commuting from their homes to their places of work in Caracas (especially after the construction of the fast train connection, *ferrocarril* between Charallave and Caracas centre), and by the still present institutional and financial incapacity of local governments to face the increasing demand for urban services (Cariola and Lacabana, 2007). Nevertheless, according to these authors, current public policies supporting strengthened citizen participation and social reforms are tackling the existing forms of socio-territorial exclusion in this area, providing the room for collective and cooperative action, and hence improvements in the living conditions of the poor.

---

1 Personal calculation based on the statistics from the 2011 National Census. Poverty calculation is realised in the 2011 National Census through the criterion of "Unsatisfied Basic Necessities". Poor households are those presenting at least one unsatisfied basic necessity, while households living in extreme poverty are those presenting at least two unsatisfied basic necessities. Basic necessities are divided into five main variables: V1: School drop-off (*Inasistencia escolar*), V2: Critical number of beds per household (*Hacinamiento critico*), V3: Inadequate housing (*Vivienda inadecuada*), V4: Lack of access to basic services (*Carencia de servicios básicos*), V5: High economic dependence (*Alta dependencia economica*). For further details, see the National Institute of Statistics website: http://www.ine.gov.ve/index.php?option=com_content&id=353&Itemid=45&view=article.

Overview of infrastructure for water delivery in the Tuy Valleys

Within the Caracas Metropolitan Region, the Hidrocapital water company is responsible for the extraction of water, its treatment, storage and distribution (this last task performed in collaboration with the state and municipal governments and, more recently, with the MTAs). The catchment area embraces the Tuy River Basin, the Guárico River Basin and the Taguaza River Basin. Water is treated in 14 stations to make it drinkable and channelled through more than 200 pump stations, which transport water to the much higher altitudes of the urban areas they supply[2]. Water service provision under the responsibility of Hidrocapital is organised into 6 water supply systems. The Losada-Ocumarito system operates in the Tuy Valleys; the other five being the Metropolitan, Coast, Panamerican, Fajardo and Barlovento. Water is supplied to the urban areas through three main pipelines called Tuy I, Tuy II and Tuy III. Tuy IV is still under construction (Martínez, 2013), connecting the Taguaza reservoir to the CMR water distribution area (Lacabana, 2003; see map 3). Many of the catchment areas, treatment stations, reservoirs and main pipes, pumping water to Caracas and to its southern periphery, fall within the Losada-Ocumarito system (Lacabana *et al.*, 2004).

With the population growing in both the Tuy Valleys and the Caracas Metropolitan Area, water production has proved to be inadequate in responding to the demand related to drinking water and the treatment of sewerage, especially in the low-income settlements. But growing urbanisation is not the only water management problem in the Tuy Valleys. Water supplies require supplementary work efforts as waste water from the residential and industrial areas is dumped into the rivers without any treatment (Lacabana, 2003; De Lisio, 2009; Martínez, 2013). Also, as Cariola and Lacabana (2007) underline, the management of water and sanitation services can also result in regional conflict. Priority over water provisions is given to the inner city supply, rather than to the Tuy Valley urban areas. But since many of the natural water resources are located in the Tuy Valleys, this sub-region provides water to the inner city while at the same time suffering from a deficit in drinking water provisions.

Hacienda el Carmen in the Paz Castillo municipality

In the Paz Castillo municipality, the population growth rate has been constant throughout the last half century (INE, 2001), but it was exceptional during the 1990s as the population has now almost doubled (De Lisio, 2009).

2 www.hidrocapital.com.gov; last accessed in 18.11.2013.

Map 3. Water infrastructure: the Tuy systems

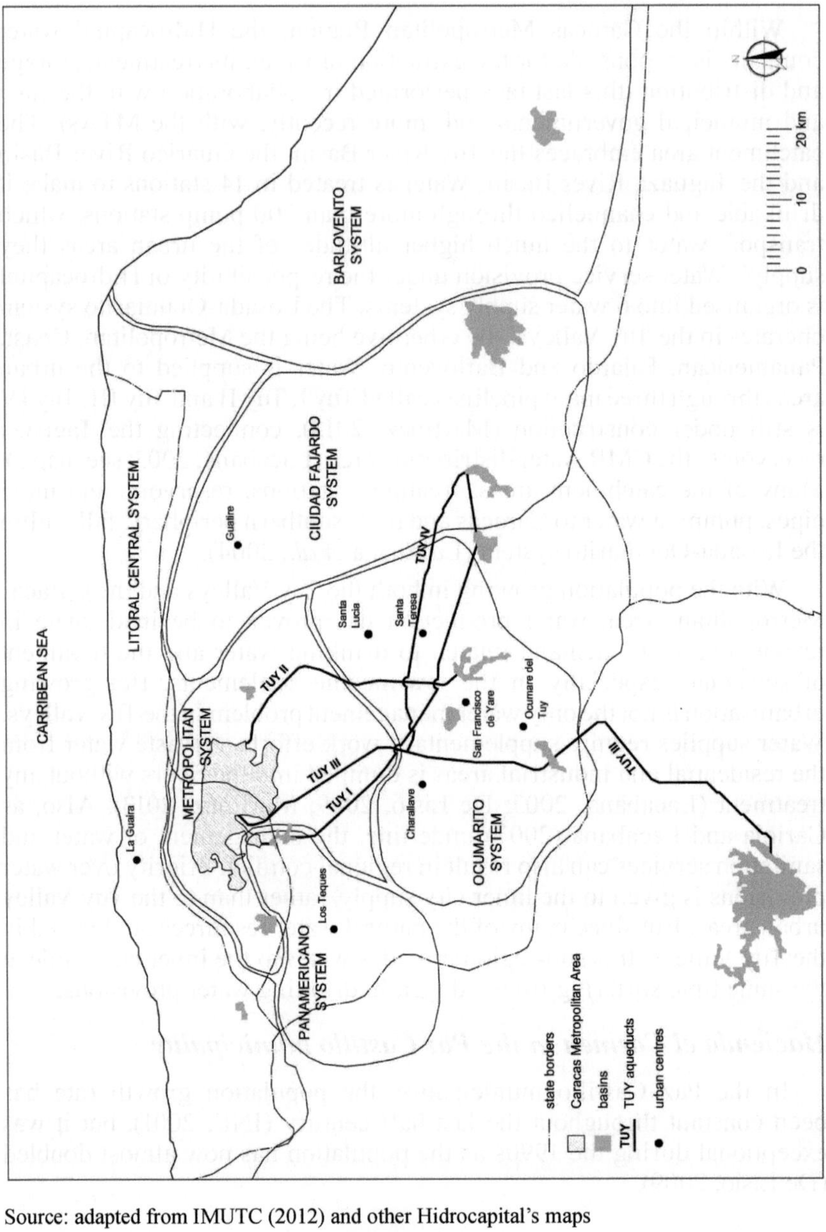

Source: adapted from IMUTC (2012) and other Hidrocapital's maps

Paz Castillo is characterised by good availability of public land which can be occupied by low-income households, by a high concentration of the poorest population sectors (also with respect to the other municipalities in the Tuy Valleys), and by the resultant presence of numerous and large informal settlements.

The *Hacienda el Carmen* settlement (HeC), located within the Paz Castillo municipality, originated from a land occupation in 1999. People invading this area came from Caracas, other areas in the Tuy Valley and, especially, from the Vargas state, as a consequence of the 1999 natural disaster. Public land belonging to the National Housing Institute (INAVI, *Instituto Nacional de Vivienda*) was occupied, forcing the institution to work out a preliminary urban development plan to rearrange the settlement plots[3]. The *Hacienda el Carmen* settlement is an extremely large informal land occupation[4] (map 5), and at the time of data collection in the field (2006) the HeC was playing host to around 2500 households. The settlement is divided into five sectors: *Araguaney Amarillo*, *Araguaney Azul*, *Colinas*, *Nueva Republica* and *El Esfuerzo* (map 6). The size of the sectors varies from one to another, as well as the dimensions of the plots and streets. In the sector *Araguaney Amarillo*, for instance, there are huge plots with a high percentage of free land, while in the sector *El Esfuerzo*, houses are built very close to each other, streets are narrow and free land is limited. Despite its illegal nature, the HeC settlement is provided with several services: a transport system linking the five sectors to the nearby *El Alto* urban centre; several small informal shops; a school (for 1500 children) and a dining hall (in the sector *Araguaney Amarillo*); a medical centre and another separate primary school (in the *Colinas* sector). Many of them have benefited from the administrative and financial support handed down by the state and national institutions.

During the early period of this settlement, access to water was obtained mainly from private tank trucks, which meant the cost of water was extremely high, a huge amount of residents' time was wasted waiting for the trucks, water quality was poor and social conflicts regularly broke out, caused by competition in accessing water[5]. As a result, HeC civil associations worked out a survey to quantify their huge expenses and, thus, lobbied Hidrocapital for access to cheaper and more reliable water provisions. In order to reduce the huge household water expenses, Hidrocapital proposed the formation of a water committee and condominiums in each sector.

3 Information collected through interviews with community leaders.

4 According to Hidrocapital officials and community residents, the HeC settlement is the largest invasion in Venezuela.

5 Residents close to the borders of the settlement – where main streets were more suitable for trucks – were advantaged in been served by water tanker trucks, for instance.

Map 4. Selected low-income communities in the Tuy Valleys

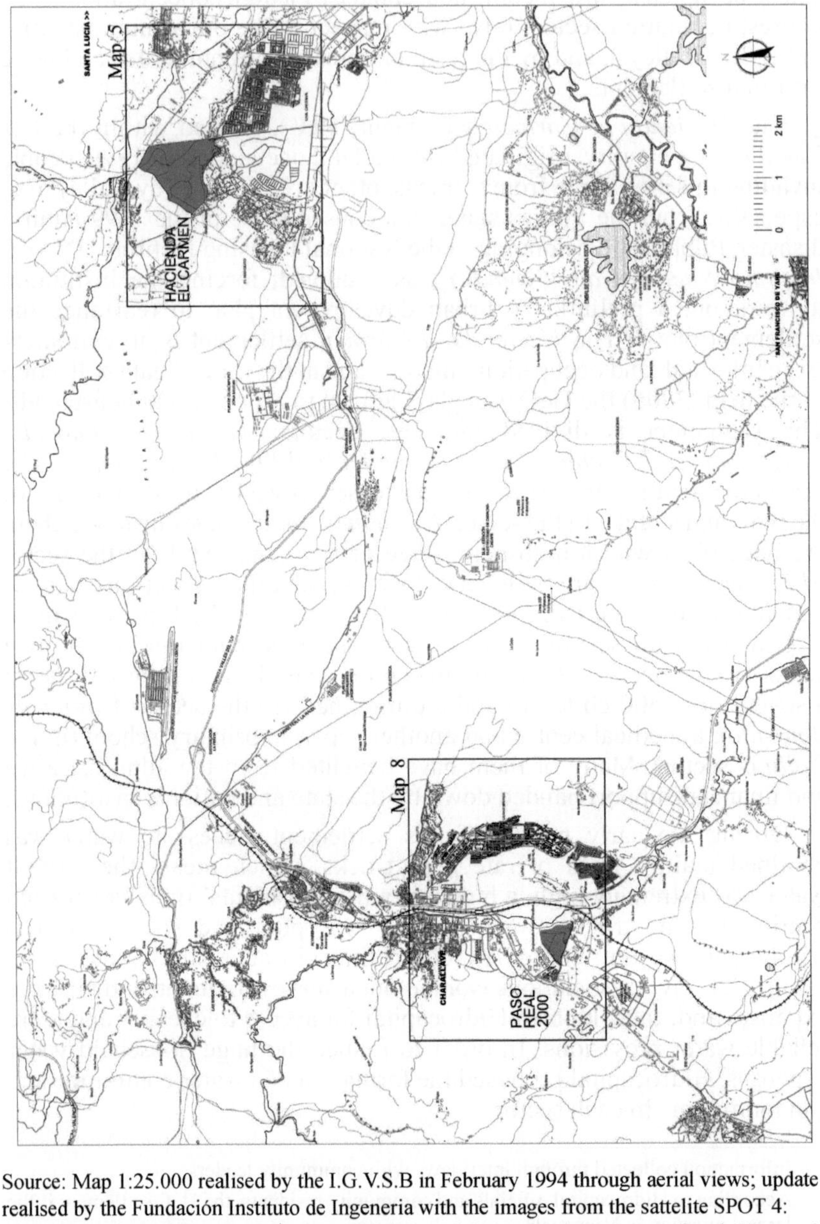

Source: Map 1:25.000 realised by the I.G.V.S.B in February 1994 through aerial views; update realised by the Fundación Instituto de Ingeneria with the images from the sattelite SPOT 4: 658-329 and 659-329 on the 31st of March 2001; aerial view made by Aerografias de Venezuela C.A. with photos taken on April 1991 (all the documents were obtained by the Cristóbal Rojas municipality).

Map 5. Hacienda el Carmen - El Cartanal

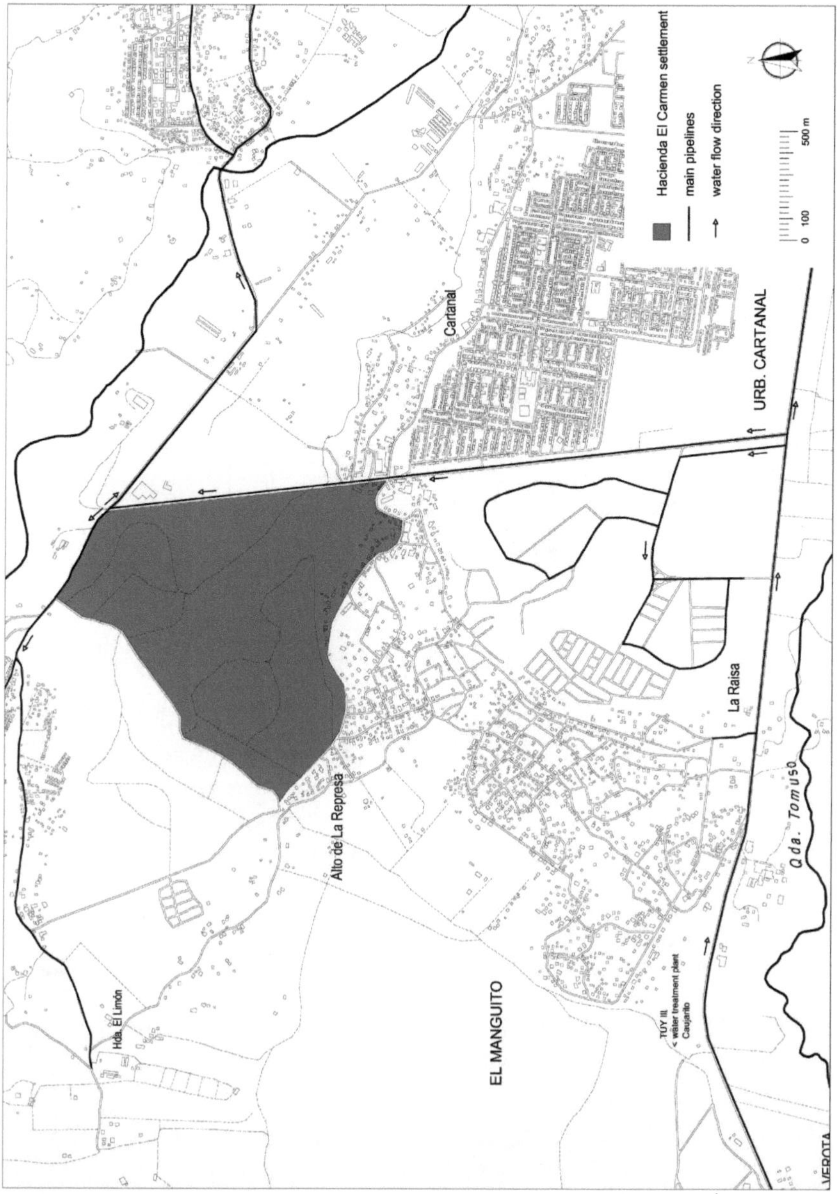

Source: adapted from the sources listed in Map 4.

Map 6. Hacienda el Carmen settlement

Source: adapted from the sources listed in Map 4 and personal elaboration.

Two different phases characterised the MTA's constitution in the HeC settlement. In the first phase, in 2003, a single water committee was set up for the whole settlement. Representatives were elected in each sector, mainly amongst the leaders already belonging to the Civil Association. However, the water board was unsuccessful due to conflicts among the leaders and a certain unresponsiveness by the Hidrocapital engineer who at the time was in charge of the HeC settlement[6]. In the second phase, another MTA was constituted in 2004, with 10 different community representatives from the various sectors and a different Hidrocapital engineer[7]. In this phase, the MTA drew up the settlement map on the basis of the INAVI plans. Afterwards, each settlement's sector was further divided into smaller condominiums to carry out the census and to collect the necessary financial resources from amongst the residents to construct the secondary pipelines. The water network was first built in the *Colinas* sector through a connection to the Tuy III (Camatagua) system's main pipe (map 3), at the end of 2004. All the other sectors were provided with a drinking water network by the end of 2005.

In 2006, the new water provision system was completed, when, through the joint work of the MTA and Hidrocapital, drinking water was made available to the whole settlement through a piped network, making the previous tanker truck deliveries redundant. The final phase in the service delivery process was the organisation of the water payment collection system. At the time of investigation, the MTA in the HeC settlement was working on the establishment of a cooperative for the whole settlement, in order to collect payments for the water service and to provide maintenance for the network.

The MTA in the HeC has been establishing tight relationships with the municipal, state (*gobernacion*) and national governments who have contributed from a financial, technical and administrative perspective in the construction of the formal water network (further details on the water project's developments are provided in chapter 6). As a consequence, the HeC's MTA has always participated in the National Meetings on drinking water and sanitation, in order to share their profound achievements with other Venezuelan MTA leaders and public institutions representatives.

6 Information collected through interviews with community leaders.

7 As the MTA in the *Hacienda el Carmen* has never been officially registered, there are different opinions with respect to the fact that there are five MTAs or just one MTA in this settlement. Latter MTA leaders agree that the current MTA is composed of 10 representatives: 3 from the sector *Araguaney Amarillo*, 1 from the sector *Araguaney Azul*, 2 from the sector *Colinas*, 2 from the sector *Nueva Republica* and 2 from the sector *El Esfuerzo*. According to the interviews with MTA leaders only 5 of them were previously involved in the Civil Associations.

Paso Real 2000 in the Cristóbal Rojas municipality

The administrative centre of the Cristóbal Rojas municipality is also its capital, Charallave, which hosts some of the specialised services belonging to the sub-region. First of all, the offices of the Tuy Valleys' *Mancomunidad de Municipios* are located in Charallave and secondly, since 2007 Charallave has been connected to Caracas by a fast metropolitan train (*ferrocarril*) which covers this distance in 14 minutes and is the fastest way to reach the metropolitan area from the Tuy Valleys. The population growth rate in Cristóbal Rojas has been constant throughout the last half century (about 14% on average), although it is increasing slower than in the Paz Castillo municipality (INE, 2001). The population is composed of both low-income and middle-income households, which lets to quite an important social heterogeneity; illegal settlements are fewer here than in the other Tuy Valleys municipalities and wide urbanisation projects for middle-income residents have been constructed (Cariola and Lacabana, 2007).

Paso Real 2000 (PR2) is the largest land occupation in the Cristóbal Rojas municipality (Lacabana, 2003), located in the southern part of Charallave (map 7). Urbanisation initially began in the area in order to create an industrial park on land belonging to a public financial institution (the Deposit Guarantee Fund – FOGADE) and a number of private companies. For this project, the area was divided into 11 terraces of irregular shape and located at different altitudes; the area was provided with streets and sidewalks together with two points of access from the two main roads around the settlement. But the industrial park was never completed and the area was abandoned until 2000, when it was invaded by spontaneous occupation. After an initial attempt to displace the households occupying the area, the municipality allowed the community to stay in the settlement. Full official recognition, however, was still pending as a consequence of the complex land tenure problem. This area in fact, continues to be subject to a long-lasting zoning dispute, which prevents the administration from resolving the land propriety rights problems in the occupiers' favour. According to the Director of "Urban and Rural Planning", the area is still classified as an industrial zone and is not residential, even if he confirmed that the community can request the change and, thus, start the process of obtaining the land rights. But according to community leaders, the area has already been recognised as residential by the municipality, so residents could already claim propriety titles.

The informal settlement organised on the basis of the 11 terraces is defined by the industrial zoning. The terraces are very different in size and densities (map 8). The plots are also of different sizes as a result of their increased subdivision. They often lack direct access from the internal roads – which creates social tensions amongst the residents due to the continual passage of people – while some of the houses are built

Map 7. Paso Real 2000 - Charallave

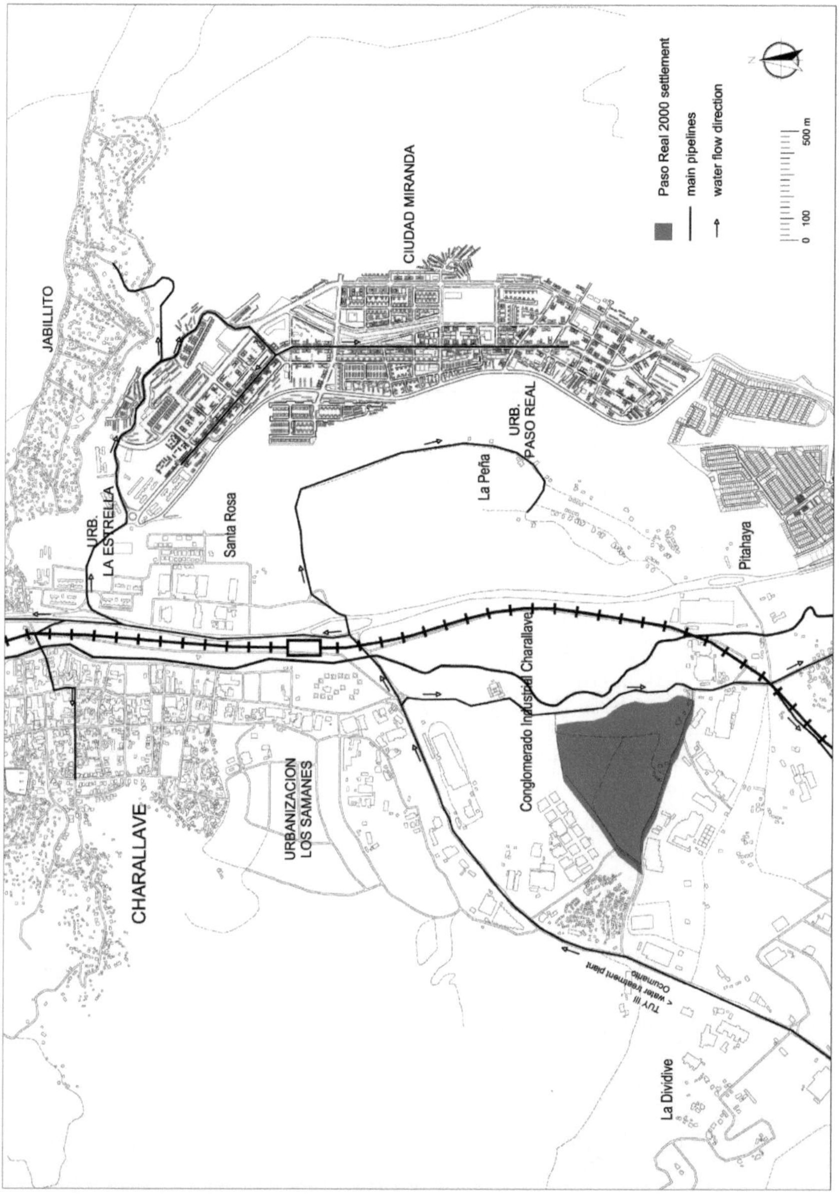

Source: adapted from the sources listed in Map 4.

Map 8. Paso Real 2000 settlement

Source: adapted from the sources listed in map 4 and personal elaboration.

on dangerously sloping areas. In 2005, about 25 houses were destroyed after a landslide following heavy rain. The population in the settlement accounted to more than 900 households in 2006.

At the time of investigation, *Paso Real 2000* had never been provided with a formal water network. Residents were accessing water through illegal taps inserted into the pipelines connected to the Tuy III system (The Cúa-Charallave mains as well as interconnections with the outlet of pumping station number 32) and also to the Tuy I through La Peñita pipeline (map 3). In the beginning, the settlement accessed water through just one tap connected to a main pipe in the upper part of the area. With the passage of time and the expansion of the settlement, a second tap was also opened in the lower part. Chapter 6 illustrates the way in which the terraces were accessing drinking water provisions.

In order to improve its access to water, the community of *Paso Real 2000* decided to set up a technical water committee in 2002. A single MTA was established for the whole settlement. Water committee members included a coordinator and representatives from all of the terraces. The history of this water board has not been straightforward. After an initial period in which residents and MTA leaders worked on the census and mapping of the settlement, the subsequent diagnostic and water project phases were abandoned. As a result, MTA members increasingly lost confidence in the project and personal interests prevailed so that the water committee fell into a sort of lethargic state. This was the case until early 2005, when the technical water committee was legally formalised and was recorded as an official community-based organisation on the municipal register. In November of the same year, there was a direct intervention from Hidrocapital following increasing pressure from the MTA leaders of one terrace that was experiencing profound water shortages. In 2006, the community of *Paso Real 2000* found itself struggling with the Urban Land Committee over management of the financial resources obtained by the central government for the housing project[8]; the MTA fell dormant for the second time, while waiting for concrete support from the municipality and Hidrocapital. Thus, household access to water still relied on the informal strategies and practices carried out by the community residents.

4.2. A Multi Level Mixed Model Design

The application of the UGI urban governance framework is carried out through a double approach that combines quantitative and qualitative

[8] The national programme for the housing sector is known as the "*substitucion de ranchos por viviendas*" (substitution of shacks with houses).

methods, through a multilevel mixed model design[9] (figure 4.1.). The use of such a combined approach aims to gain insight into the formal and more institutional dimension of urban governance at the municipal level, as well as into the more informal, local and community-based aspects of governance arrangements for water services, without relying exclusively on the information given by administrative official data. This amendment is also supported by the fact that the Campaign on Urban Governance foresees the possibility of complementing the core set of quantitative data with more qualitative surveys (UN-HABITAT, 2004a).

Figure 4.1. Multilevel Mixed Model Design

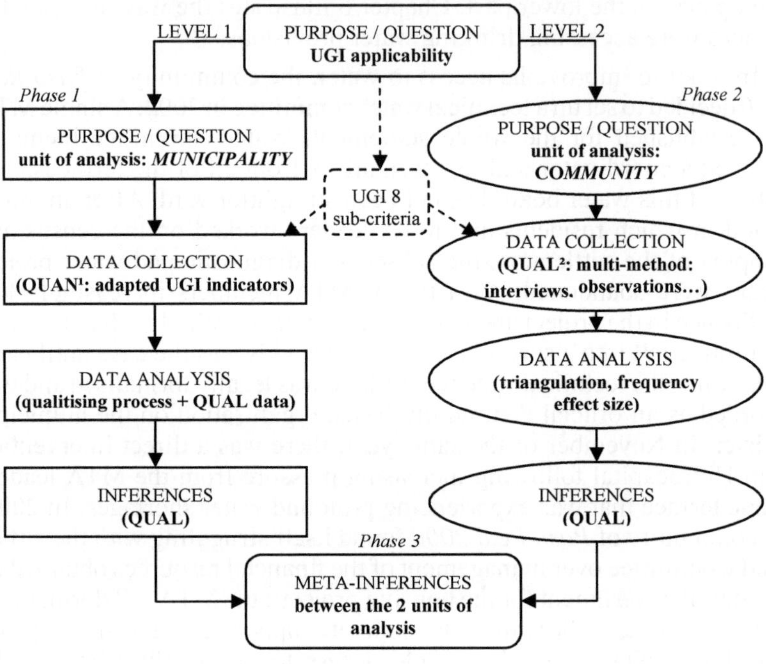

Source: adapted from Tashakkori and Teddlie (2003).
[1] QUAN: indicates a quantitatively-driven project (Morse, 2003).
[2] QUAL: indicates a qualitatively-driven project (Morse, 2003).

---

9 According to Tashakkori and Teddlie (2003: 711), a multilevel mixed model design "is a design in which QUAL data are collected at one level (e.g., child) and QUAN data are collected at another level (e.g., family) in a concurrent or sequential manner to answer interrelated research questions with multiple approaches (QUAL and QUAL) that are pulled together at the end of the study in the form of 'global inferences'". The *mixed model* design relies on the mixing of QUAN and QUAL approaches at several stages of the study where, conversely, a *mixed method* design involves the mixing of QUAN and QUAL approaches only at the methods stage of a study.

Urban governance assessment can thus be organised into three different phases. The first two phases relate to the assessment of urban governance arrangements at municipality and community levels respectively, which represent the two units of analysis. The third phase concerns the comparison of inferences stemming from these two levels of analysis. Between the 4 UGI core principles and the 25 UGI indicators, 8 sub-criteria have been identified according to the definition of the principles of effectiveness, equity, participation and accountability provided by UN-HABITAT, in order to define a common analysis framework amongst the three different phases. Data collection questions, data analysis, inferences and meta-inferences are all organised on the basis of the following eight sub-criteria:

PRINCIPLES	SUB-CRITERIA	
effectiveness	Sub-criterion 1:	Subsidiarity of authority and sufficient resources and autonomy
	Sub-criterion 2:	(Institutional) efficiency in delivering public services and responding to civil society concerns and welfare
equity	Sub-criterion 3:	Distributional equity
	Sub-criterion 4:	Procedural equity
participation	Sub-criterion 5:	Representative democracy
	Sub-criterion 6:	Participative democracy
accountability	Sub-criterion 7:	Downward Accountability
	Sub-criterion 8:	Upward Accountability

Before moving on to the methodological description of these three phases, it has to be remembered that the Index was first adapted to the context of urban water supply as an example of basic service around which alternative governance arrangements develop to guarantee improved access. All the indicators and their adaptations to the community level of analysis (*phase 2*) have focused on drinking water, when this adjustment was possible[10].

Phase 1: Municipal level of analysis

This strand aims at exploring and understanding the extent and the quality of governance at the municipal level in improving water supplies (including in low-income communities), by applying the quantitative Urban Governance Index indicators. Factual data and information for calculating the UGI indicators in the two selected municipalities was collected in two ways:

10 UGI indicators' adaptations to the water sector have been done solely for those indicators concerning basic services in general. In a first phase, we tried to adapt the full twenty-five indicators to the water sector but, for many of them, this change was not possible or, where possible, data was often not available at the municipal level.

- direct consultation of, first, the official national statistics elaborated by national state bodies – notably, the *Instituto Nacional de Estatistica* (INE) and the *Consejo Nacional Electoral* – and generally available through access to the respective websites[11], and, second, the municipal budgets and accounts available at the time of the field investigation (2006)[12];
- interviews with local government and state-owned water company officials. In this way – according to UN-HABITAT (2004a) recommendations – the quantitative data has almost always been verified through urban experts and urban policy makers involved in city planning.

The quantitative data collected through the completion of the UGI indicators is also used to explore local governance in the municipalities of Paz Castillo and Cristóbal Rojas, through a "qualitising process", by which quantitative data is transformed into, and then analysed as, qualitative information (Tashakkori and Teddlie, 2003, 2010). This process aims at drawing "verbal portraits" (Sandelowski, 2000) or "narrative profiles" (Tashakkori and Teddlie, 1998) concerning the specific target phenomena and based on the quantitative data previously collected. In general, "qualitising can be used to extract more information from quantitative data or to confirm interpretations of it" (Sandelowski, 2000: 253). Actually, the qualitising process applied in this research helped to reveal aspects of the local governance arrangements that the UGI indicators did not.

A qualitising process is applied to *phase 1* in order to explain the significance of the UGI indicators' data for the two selected municipalities and to confirm or refute the Venezuelan general trend regarding urban governance at the municipal level (focusing on water provisions). In this phase, the unit of analysis is represented by the two selected municipalities; the initial quantitative data are the UGI indicators, and the target phenomena under analysis are represented by the 8 UGI sub-criteria for which as many profiles are drawn. Amongst the main profile types (modal, average, holistic, comparative and normative; Tashakkori and Teddlie, 1998), this analysis relies mainly on a comparative profile[13], where comparison is made against the best score for each indicator.

---

11 http://www.ine.gov.ve and http://www.cne.gov.ve.

12 To ensure coherence between the municipal data in the budgets and accounts collected from the two municipalities and other quantitative data available through national statistical databases providing more recent information (such as the percentage of households accessing or voter turnout), all data included in the UGI calculation is from the year 2006 or before. When applicable, essential following changes, with respect to the situations portrayed up to this date, are presented in the text (see mainly chapter 5).

13 A comparative profile is a narrative description based on the comparison of cases to each other on one or more sets of scores (Tashakkori and Teddlie, 1998).

Table 4.1. Indicators belonging to the different UGI core criteria

UGI Indicators		UGI core criteria			
		EFFECT.	*EQUITY*	*PARTICIP.*	*ACCOUNT.*
1	Local Government revenue per capita	●			○
2	Ratio of recurrent and capital budget	●			○
3	Ratio of mandated to actual tax collected	●		○	○
4	Local government revenue transfer	●		○	○
5	Predictability of transfers in local government budget	●			
6	Published performance delivery standards for water supply	●			○
7	Consumer satisfaction survey for water services	●			○
8	Vision statement effective	●		○	○
9	Citizens charter for water services		●		○
10	Percentage of women councillors		●		
11	Percentage of women in key positions		●		
12	Pro-poor pricing policy for water		●		
13	Incentives for informal arrangements to improve water access		●		
14	Elected council	○		●	
15	Locally elected Mayor	○		●	○
16	Voter turnout	○	○	●	○
17	Public forum for water			●	
18	Civic associations per 10,000 pop	○	○	●	○
19	Formal Publication	○	○	○	●
20	Control by higher Govt.	○		○	●
21	Codes of conduct		○	○	●
22	Facilities to receive complaints	○		○	●
23	Anti-corruption commission		○	○	●
24	Personal Income and assets				●
25	Regular independent audit				●

● = core criterion to which the indicator principally belongs to
○ = core criterion to which the indicator belongs to, but to a minor extent
Source: adapted from UN-HABITAT (2004c).

The qualitising technique applied to the UGI indicators in order to develop sub-criteria profiles employed three additional and qualitative data sources which contributed to the creation of the narrative profiles. First, UN-HABITAT concept papers and reports in order to investigate the significance of the sub-criteria; second, relevant scientific literature about the indicators and sub-criteria themes – especially in the context of Venezuela – in order to outline the hypothesis to be verified in the two municipalities; third, interviews with key actors belonging to the two municipalities, Hidrocapital, Hidroven and academia in order to integrate the quantitative indicators with further qualitative information. In addition, given the possibility of overlapping between sub-criteria (see for instance, "procedural equity" and "participative democracy") and of indicators' belonging to different criteria and sub-criteria, outcomes stemming from the elaboration of each sub-criterion profile are continuously integrated in the other profiles. Table 4.1. illustrates how each UGI indicator belongs primarily to one of the four UGI core criteria but, very often, also to the others.

Phase 2: Community level of analysis[14]

The second phase aims at exploring how the four basic UGI principles can be translated into a qualitative exploration of the informal side and community dynamics of local governance systems in improving access to urban basic services. This phase intends to examine the degree of involvement and inclusion of low-income communities in these governance arrangements through technical water committees (*mesas tecnicas de agua* – MTA). In other words, qualitative data is used to investigate how the urban governance principles framework works in practice. Contrary to the previous phase, this part of the analysis is based on qualitative information, which attempts to highlight the communities' perceptions of and level of involvement in these local governance systems.

The UGI principles were tested in the two previously described low-income settlements located in the Middle Tuy Valleys. The two settlements were selected as a "matched sample" (Laws *et al.*, 2003) due to the presence of a technical water committee in each settlement and a similar period of constitution (around 1999). They thus offered the possibility to investigate two different advancements of MTA activities. In *Paso Real 2000* (Municipality of Cristóbal Rojas) the MTA was set up a few years after the community was established, although it faced numerous problems that slowed down its activity and affected improvements in water supplies. Conversely, in *Hacienda el Carmen* (Municipality of Paz Castillo) the MTA was set up later, the service was improved significantly and the community representatives were well-organised enough to improve water supply.

---

14 Fieldwork strongly benefited from previous research and contacts established by the DPU (Development Planning Unit, University College London) and the CENDES (Centro de Estudio del Desarrollo, Universidad Central de Venezuela) in the framework of the research project "Service provision governance in the peri-urban interface of metropolitan areas", funded by the DFID.

The unit of analysis in this phase is represented by these low-income communities, but given the considerable size of the settlements, a specific area was chosen in both of them. In *Paso Real 2000*, the analysis was carried out in *terraza 11*, which is the most densely populated *terraza* of this settlement (map 9 and 10). This sector was selected for two main reasons: first, the presence of a recent water pipeline project realised by the state-owned company in collaboration with the MTA; second, security reasons suggested not to conduct the analysis in the higher part of the settlement. In *Hacienda el Carmen*, two out of its five sectors were selected to run the analysis: *Araguaney Amarillo* and *Colinas* (map 6). Our choice was based on three motivations. First, *Colinas* was the only sector in the *Hacienda el Carmen* where households were already paying for the water service, whilst the others were about to start paying; second, the MTA leaders who could introduce us to the settlement belonged to these two sectors; finally, these two sectors border on each other, making it easier to run our interviews.

Interviews represented the primary source of qualitative information collection for the fieldwork. Semi-structured interviews were carried out in order to gain qualitative and, to a minor extent, quantitative information about the local governance arrangements for accessing water supply. According to the diverse groups of interviewees directly involved in the process of service delivery (the municipality, the water company, the water committees and community residents)[15], different question sets were designed on the basis of the qualitative applicability of the principle framework of the Urban Governance Index (see previous chapter). Table 4.2. provides the list of questions that were asked to all selected interviewees. Question selection was made in the light of two main issues. First, the number of questions: despite the complexity of these principles, we were obliged to select only a few representative questions – “the absolutely necessary questions” (Laws *et al.*, 2003) – so as not to slow down the interview and take too much time. Second, as with the UGI indicators, some questions might belong to more than one specific principle (see, for instance, how UN-HABITAT, 2003c, relates each UGI indicator to diverse principles). The respondents’ answers are thus used in a flexible way across the different sub-principles.

In the case of community members, interviews were generally conducted through repeated one-to-one meetings, often run in an informal manner, benefiting from – but at the same time sometimes suffering from – the different levels of people’s confidence regarding our research. In general, in each sector, semi-structured interviews were run in households chosen through a “purposive quota sample” (Laws *et al.*, 2003) in order

[15] The complete list of all persons interviewed during the fieldwork is present in Annex 1.

Map 9. Hacienda El Carmen surveyed households

Source: personal elaboration.

Map 10. Paso Real 2000 surveyed households

Source: personal elaboration.

to gain insight into the different population groups and trying to rely on a balanced representation for each group. In *Paso Real 2000*, within *terraza 11*, interviews were run in 24 households, trying to balance the representation of households benefiting from the water company project with those still obtaining water through illegal connections (by the new project or the main pipe; map 10). Nevertheless, the selection was heavily influenced by the repeated absence of any family member during the daytime. For the *Hacienda el Carmen*, 12 households were interviewed in *Colinas* and 11 in *Araguaney Amarillo* (map 9). These community members were provided with the same water supply system but were experiencing different payment conditions, according to the sector where they were living[16]. The number of interviews realised with the community members cannot represent a statistical sample. Nevertheless, it is possible to identify a "frequency effect size" (Onwuegbuzie and Teddlie, 2003) related to the rate of the prevalence of certain types of answers given by community members[17]. This operation can also benefit from the option to "binarise" the answers given by residents into a "yes" or "no" answer (see Annex 2). Results from interviews with dwellers are, however, always triangulated with other sources of evidence as explained below.

Semi-structured interviews with MTA leaders, municipal and water company public officials took between 30 minutes and one hour and were generally taped. Other pieces of information have also been collected amongst these groups of interviewees through a more open scheme, which benefited from many informal contacts and exchanges. For example, casual dialogue with MTA leaders often occurred during our stay in the communities. These circumstances represented, for instance, an opportunity to discover more about the history of the settlement as well as the presence and activities of many other CBOs. Also, car journeys with Hidrocapital officials often gave us the opportunity to informally speak about the water company's strategies for dealing with the MTAs.

The selection of institutional interviewees (belonging to the municipality and water company) was done according to three main criteria. First, key persons involved in the water delivery process were identified with the help of local academicians and researchers[18]. Second,

16 The choice of households with whom to hold the interviews was strongly conditioned by the presence of the community leaders who accompanied us during the interview process in the *Hacienda el Carmen*, in order to prevent security problems.

17 According to Onwuegbuzie and Teddlie (2003), frequency effect size refers to the frequency of a particular theme within a sample and it can be converted to a percentage.

18 Professor Miguel Lacabana – based at the CENDES – *Centro de Estudio del Desarollo* of the Central University of Venezuela – was our main source here, because of his extensive research on MTAs in Venezuela, but also Marianela Carrillo and Cecilia Cariola, also at CENDES, supported our interviewee selection.

other key persons had been identified by the municipal and water company public officials, according to the snowball sampling technique. Finally, some interviewees were chosen because of their relevant position in the service delivery chain. For instance, at the municipal level, we spoke with the directors (or similar positions) of the departments dealing with "Public Infrastructure", "Urban conservation and maintenance", "Urbanism and Infrastructure" and "Public Services", just to cite a few of the most relevant departments.

During the interviews with community residents, MTA leaders, public servants belonging to the two Municipalities and state-owned water company (Hidrocapital) officials, a scoring exercise was also proposed on an individual basis in order to assess the level of contribution and collaboration amongst the main local stakeholders involved in the water supply delivery process. It is interesting to note that this ranking exercise provides a measure of both effectiveness and participation at the local level (see box 4.1.). Finally, a different, open interview scheme was designed to involve the representatives of the main international aid agencies in Caracas (the World Bank, the UNDP and the Delegation of the European Commission in Venezuela). The purpose of these meetings was to understand whether and how these organisations were involved in the local governance arrangements to improve access to urban water supplies, and also to obtain an external view of the national changes occurring in the water and sanitation sector.

Box 4.1. Scoring exercise

This scoring exercise has been designed to understand what value local key stakeholders would give to their effective contribution and other stakeholders' contribution to improving water supply, for the chosen low-income communities. Community members, MTA leaders, public servants belonging to the two municipalities and state-owned water company (Hidrocapital) officials were asked to weight these contributions (from 1 to 5, where 1 = very low and 5 = very high).

	1 = very low	**2**	**3**	**4**	**5 = very high**
community members					
MTA					
Municipality					
Hidrocapital					

To conclude, in order to triangulate the information collected through the interviews, the fieldwork investigation was carried out using multiple sources of evidence, such as documentation collected at the local level, participant and direct (or systematic) observation, focus groups and physical artefacts.

Table 4.2. Semi-structured interview framework

	community members	water board leaders	water board focus group	municipal public servants	state-owned water company officials
effectiveness					
effective delivery of public services	– Do you think that the MTA has been important in improving your access to water?	– Up to now, how do you think that the MTA has contributed to improving water access in the community? – How do you think the MTA could improve the current water supply for the whole community?	– What are the results of and changes in MTA activities in the community? – What are the limitations, difficulties and problems in the MTA activities?	– What are the results of and changes in MTA activities in the community?	– What are the results of and changes in MTA activities in the community?
responsiveness to civil society concerns	– Did any of the MTA leaders ask you about your water shortage?	– How does the MTA become aware of community water related problems?	– What could the MTA do to improve the water supply for the community?	– How does the municipality become aware of community water related problems and MTA activities?	– How does Hidrocapital become aware of community water related problems and MTA activities?
equity					
access to basic services	– Up to now, how has your access to water improved? – How do you think that the MTA could improve your current access?	– What is the purpose of the MTA?	– What is the purpose of the MTA?	– What is the municipal's aim with regards to water supply for the (selected) community?	– What is Hidrocapital's purpose with regards to water supply for the (selected) community?

	community members	water board leaders	water board focus group	municipal public servants	state-owned water company officials
equity in decision making	– What do you think about water payment? – Do you pay for water? – Did you agree to pay for water? – Did you share your opinion with an MTA leader?	– Was there a meeting to speak about water payment? – How much does the community pay for water? – Did the MTA ask the community about the water payment?	– How do you think the water payment should be organised? – How did you speak about the water supply payment with the community and with Hidrocapital?	– How does the municipality involve the community/MTA in the decision-making process? – What is the municipal position with regards to water subsidies?	– How does Hidrocapital involve the community/MTA in the decision-making process? – How is the tariff structure organised?
participation					
participative democracy	– Do you know the MTA? – How did you hear about the MTA? – Have you ever spoken about your water related problems with an MTA leader? – Did you collaborate at the pipeline construction?	– How many people participate at the MTA meetings? – What are the relationships with the Municipality and Hidrocapital? – Is there a 'water community council'?	– What are the relationships with the community, the other community based association, the Municipality and Hidrocapital?	– What are the relationships with the MTA? – What are the relationships with Hidrocapital?	– What are the relationships with the MTA? – What are the relationships with the municipality?
representative democracy	– Were you aware of the meeting in which the MTA leaders were elected?	– How were the MTA leaders elected?	– How were the MTA leaders elected?	– How are the MTA leaders chosen?	– How are the MTA leaders chosen?

	community members	water board leaders	water board focus group	municipal public servants	state-owned water company officials
accountability					
transparency	– Do you know the purpose of the MTA? – Do you know the outcomes of the MTA meeting?	– How does the MTA inform the community about meetings' with the municipality and Hidrocapital?	– How does the MTA inform the other stakeholders about the changes and decision about water supply?	– What are the municipal responsibilities with regards to water supply? – How does the Municipality inform the community/ MTA about its responsibilities?	– What are Hidrocapital responsibilities with regard to water supply? – How does Hidrocapital inform the community/ MTA about its responsibilities?
integrity		– Is there a MTA written regulation?	– What and how are the relationships with the other community-based committees (as the CTU)?	(*UGI* quantitative *indicators)*	(*UGI* quantitative *indicators)*
responsiveness	– Did any of the MTA leaders ask you about your water shortage?	– How does the MTA become aware of community water related problems?		– How often do you meet the MTA leaders? – Is there an office where people can complain?	– How often do you meet the MTA leaders? – Is there an office where people can complain?
willingness to take and share responsibilities	– Are you willing to be actively involved in the pipeline construction and in the MTA meetings? – Did you collaborate in the pipeline construction?				

Phase 3: Meta-inferences from the two units of analysis

This methodological approach leads to a comparison between the inferences resulting from the previous two phases (quantitative UGI application at the municipal level and the respective qualitative investigation with the selected communities). The analysis in this phase aims at understanding, in an integrated way, the two facets of urban governance for service co-production: the formal, more institutional dimension of urban governance at the municipal level, as well as the more informal, local and community-based aspects. The integration of quantitative (municipal level) and qualitative (community level) approaches may provide three different outcomes, as a comparison of results can be convergent, divergent or complementary (Erzberger and Kelle, 2003). Meta-inferences from the two levels and units of analysis are worked out according to a narrative integration and comparison of the "qualitised" inferences belonging to *phase 1* and the qualitative results belonging to *phase 2* of the study.

This stage of analysis primarily rests on a "mutual validation model", which searches "for convergent findings as the most important purpose" (*ibid.*: 484) of the comparison between the municipal and community levels of analysis. At the same time, it also foresees a "complementary model", which "draws our attention to the fact that different methods may relate to different aspects of the domain in question because certain methods have been developed within different theoretical traditions and thus serve the investigation of different phenomena" (*ibid.*). The measurement of urban governance through quantitative indicators and their more statistical and factual calculation, and the assessment of local governance arrangements and their habitually informal and unquantifiable relationships fall into this idea. Both models, however, "require a common theoretical framework for the employment of qualitative and quantitative data collection and analysis" (*ibid.*), which is actually provided by the Urban Governance index principle framework.

CHAPTER 5

Assessing Urban Governance for Water Services at the Municipal Level

5.1. Poor effectiveness within an incomplete decentralisation process

Institutional efficiency at the municipal level will be assessed against the two previously introduced effectiveness sub-criteria. One is formed of subsidiarity of authority, availability of resources and the autonomy to meet responsibilities and manage revenue resources (with a specific focus on water provisions), and the other of the mechanisms in place for the effective delivery of water provisions and responsiveness to civil society concerns (vision statement). Assessing these local government characteristics inevitably implies entering into a complex and sensitive evaluation of the decentralisation processes underway in Venezuela, from a political, legal, administrative and fiscal perspective.

Calculating the UGI indicators in Paz Castillo and Cristóbal Rojas reveals poor municipal effectiveness in both local governments (see table 5.1.). Poor performance depends on the ratio of recurrent and capital budget (UGI indicator No. 2) and on the local government revenues from transfers (UGI indicator No. 3), in which they equally score very low, together with a zero score for the UGI indicators linked to the provision of water supplies (No. 6 and 7). In particular, the effectiveness sub-criterion of institutional efficiency in delivering water services and responding to civil society concerns and welfare scores zero in all of the three specific indicators (UGI indicators No. 6, 7 and 8, with a partial exception for indicator No. 8 in the Cristóbal Rojas municipality). Nevertheless, despite these fairly low outcomes, results for the local government revenues per capita (UGI indicator No. 1) are over the average of the UN-HABITAT second field test (UN-HABITAT, 2004) and transfers to local government budgets seem correctly predictable (UGI indicator No. 5).

Elaboration of the qualitative narrative profiles for the two sub-criteria (§ 5.1.1. and 5.1.2.) helps to explain why the effectiveness sub-indexes score so low in both municipalities, whilst confirming the hypothesis, supported by much critical literature, that the decentralisation process currently taking place in Venezuela has been sustained and implemented at a political and normative level but not at a fiscal and administrative one. It seems possible to argue that the municipalities of Cristóbal Rojas and Paz Castillo encounter the same problems faced by the other Venezuelan municipalities in managing local government affairs in a decentralised manner. Many critics contend that the uncertainty linked to the blurred roles and responsibilities that municipalities hold in the delivery of public services and the insufficiency and rigidity of the financial resource allocation framework show that priority has been given to the political viability of the decentralisation process rather than to the proposal's technical execution (Mascareño, 1996, 2003; Rojas, 2000; Leal Rìos and Gotera, 2003; McCoy, 2004; D'Elia, 2010). Other critical literature underlines how the supposed decentralisation process is steeped in a strong centralised trend, characterising the current Venezuelan institutional structure and thus hampering a full and comprehensive managerial and administrative decentralisation (Mascareño, 2006; Borgucci, 2003, 2013; de la Cruz, 2004; Lalander and Samaniego, 2005; Rachadell, 2006; Cuñarro *et al.*, 2006; Brewer-Carías, 2010; D'Elia, 2010; Penfold, 2010; Smilde, 2010; Córdoba Jaimes, 2012; Peraza, 2012).

The decentralisation of water service provisions to the two selected municipalities is also facing major difficulties because of an incomplete fiscal and administrative decentralisation. Mechanisms required for the efficient delivery of water services (such as a published performance delivery standard) and those indicating the willingness of the local administration to receive critical feedback from the local population (such as a customer satisfaction survey) cannot in fact be found at the municipal level. It can therefore be asserted that there is evidence supporting the hypothesis that the poor effectiveness of the two municipalities in local governance rests much more on shortcomings linked to the current decentralisation process at the national level, rather than on the municipality's supposedly inherent inefficiency to deliver water services, or to respond to civil society concerns with respect to these services.

Inconsistencies in the wider Venezuelan decentralisation discourse, together with the delay in responsibility being transferred from HIDROVEN to municipalities (scheduled in theory for the end of 2013; LOPSAPS, 2007) and the continued absence of new national water management bodies, contribute to prolonging the absence of a decentralised water services' management and provision operated by the municipality. In other words, the low score in sub-criteria 1 can be seen as the cause for the low scoring in sub-criteria 2.

Table 5.1. UGI effectiveness indicators and effectiveness sub-criteria

EFFECTIVENESS UGI INDICATORS				*EFFECTIVENESS SUB-CRITERIA*	
No.	**Indicator**	**Data (X)**		***Subsidiarity of authority, availability of resources and autonomy***	***Institutional efficiency in service delivery, responsiveness to civil society concerns***
		Paz Castillo	**Cristóbal Rojas**		
1	Local Government revenue per capita (LGR)	US$ 86.08	US$ 123.9	●	○
2	Ratio of recurrent and capital budget (RRC)	0.16	1.25	●	○
3	Local government revenue transfer (LGT)	86.0%	44.4%	●	○
4	Ratio of mandated to actual tax collected (TC)	*n. a.*	ca. 50%	●	○ / ●
5	Predictability of transfers in local government budget (PoT)	Yes	Yes	●	○
6	Published performance delivery standards (PPDS) for water supply	No	No	○	●
7	Consumer satisfaction survey (CSS) for water supply	No	No	○	●
8	Vision statement effective (VSE)	No	Yes/No	○ / ●	●
	Effectiveness sub-index	**0.31**	**0.49**		

● = indicator considered belonging to the sub-criterion
○ = indicator considered not belonging to the sub-criterion

Subsidiarity of authority and availability of sufficient resources and autonomy (decentralisation process)[1]

Political and legal aspects

According to both the Constitution of 1999 (art. 168) and 1961, Venezuelan municipalities are considered the primary political unit of the national organisation, with legal personality and autonomy within the limits defined by the constitution and the law. Venezuelan municipal autonomy affects three essential aspects of the decentralisation process: the election of mayors and councillors (political autonomy), the management of activities falling under municipal responsibility (administrative and normative autonomy) and the creation, collection and investment of municipal revenues (fiscal autonomy) (Mascareño, 2003; Leal Rìos and Gotera, 2003). From a legal perspective, Mascareño (2003) identifies two main waves of juridical instruments that characterised the decentralisation process that took place in Venezuela between 1990 and 1999[2]. Added to this, two more recent steps in the Venezuelan decentralisation process are found in the 1999 Constitution and the Organic Municipal Public Power Act (*Ley Organica del Poder Publico Municipal* – LOPPM) of 2005, further modified in 2010, which replaces the previous Organic Municipal Government Act of 1989. In

[1] It is worth underlying that the UGI indicators concerning the "control by higher levels of government" (UGI indicator No. 20) and the "predictability of transfers in local government budget" (UGI indicator No. 5) are both included in the check-list indicator No. 7, concerning decentralization, of the Habitat Agenda Urban Indicators (UN-HABITAT, 2004b: 50) which aims at determining the "effective decentralisation of responsibilities, policy management, decision-making authority and sufficient resources".

[2] Four specific laws belong to the first wave (1988-1989): Election and Removal of Governors Act (*Ley sobre la election y remocion de los gobernadores de Estado*); States' Public Powers Duration Act (*Ley sobre el periodo de los poderes publicos de los Estados*); Organic Municipal Government Act (*Ley organica de Regimen Municipal*); Organic Decentralisation, Delimitation and Transfer of Public Power Responsibilities Act (*Ley organica de descentralizacion, delimitacion y transferencia de competencias del poder publico*). The second phase of juridical decentralisation is characterised by two other laws approved in 1993: Decree-law for the mechanisms regulating states and municipalities participatory mechanisms for value added tax and intergovernmental tax for decentralisation (*Decreto-ley que regula los mecanismos de participation de los Estados y municipaios en el producto del impuesto al valor agregado y el fondo intergubernamental para la descentralizacion*); Special economic allocations for states derived from mining and hydrocarbons (*Ley de asignaciones economicas especiales para los Estados derivadas de minas y hidrocarburos*) Mascareño (2003).

particular, these two laws confirm the autonomy held by municipalities in the management of their responsibilities, their legal personality and their primacy as a political unit.

Nevertheless, many critics contend that the political and juridical framework set up to support the decentralisation process does not seem to be complemented by a similarly well structured scheme for effective implementation with respect to resources and responsibilities. Scholars underline that, although the macro-economic crisis – linked to reduced oil income during the 1970s and 1980s – and the resulting crisis in the central state government was undoubtedly one of the main reasons giving rise to decentralisation[3], the process has more of a political origin (Rojas, 2000; Leal Rìos and Gotera, 2003; Gonzales Villar, 2004; Brewer-Carías, 2010; Rìos *et al.*, 2012). It stems from crisis within the political parties and the resulting necessity to entrust the political system with a new legitimacy and credibility, and hence with new representative forms. So, these authors support the idea – already expressed by Knoop (1994) – that, in the Venezuelan decentralisation process, priority has been given to the political sustainability of the practice. This taking precedence over defining technical proposals to ensure abundant resources and clear responsibility division amongst the institutional actors – without overlaps, confusion and asymmetries. The analysis conducted in the municipalities of Cristóbal Rojas and Paz Castillo seems to confirm this hypothesis and is detailed below.

Administrative aspect

The decentralisation of public service responsibilities to the local governments seems incomplete and unclear. The Constitution of 1999 (art. 164) establishes that states are responsible for the "creation, management and organisation of the state public services", including drinking water. In addition, according to art. 178 of the Constitution (1999) and art. 56 of the LOPPM (2005), municipalities hold direct responsibilities in the administration of their own interests and in the management of their local life. This last point includes the provision of domiciliary public services and the improvement of community life conditions concerning urban mobility, public events, environmental protection, health, drinking water, power, justice and peace, etc. In the case of concurrent responsibilities for the provision of public services amongst municipalities

3 The new decentralised model was strongly recommended, in Venezuela also, by external multilateral financial organisations in order to achieve macroeconomic stability.

and the state or national powers, art. 57 of the LOPPM (2005) and art. 165 of the Constitution (1999) declare that these responsibilities shall be regulated according to the principles of "interdependency, co-ordination, co-operation, co-responsibility and subsidiarity" and that responsibilities that can be assumed by municipalities have to be transferred from the state power. But responsibilities between different government levels are not clearly assigned. A study of the Venezuelan decentralisation process[4], for instance, highlighted that 42% of Venezuelan mayors interviewed considered that "clear definition of local government responsibilities" is the most important characteristic of an "ideal" decentralisation policy, while stressing its current lack in their country (Reveron Escobar, 2003: 207).

It is unrealistic not to see a cause-effect relationship between the tendency towards political centralisation at the national level and the problem of blurred responsibilities at the local level. Responsibilities between Venezuelan municipalities and states appear, for instance, not to be shared out on the basis of the local government's capacity to assume further tasks (Leal Rìos and Gotera, 2003) but rather on the political project to directly involve citizens in the decision and organisation of public affairs – the "Bolivarian Project". Some scholars have recently highlighted how the Communal Councils (or Community Councils – *Consejos Comunales*) are playing, within this supposed re-centralisation process, a critical role in increasingly enhancing concurrent responsibility at the local level: on the one side, Communal Councils, by profiting from their general responsibilities and direct relationships with the national executive power, are replacing municipalities in providing services (D'Elia, 2010; Peraza, 2012) and, on the other, they also benefit from direct national funds, such as the Fund for Inter-territorial Cooperation (see following pages), which partially reduces the financial autonomy of local governments (Moreno, 2008; Rìos *et al.*, 2012). In the end, responsibilities at the local level are not only more de-concentrated than decentralised because of the centre-based administrative organisation (Borgucci, 2003, 2013), but they are also undermined, in local administrations, by the widest discretional authority of the national executive power to withdraw them from states and municipalities for strategic reasons (Peraza, 2012).

In the case of the water sector, there seems to be the same problem of concurrent sharing of responsibilities amongst different levels

4 This study included a sample of 89 mayors out of the total 334. In reality, despite her interest in "comparing what is currently achieved and what should be" in the Venezuelan decentralisation process, Reveron Escobar (2003: 207) avoids making an explicit link between the necessity for and the lack of clear decentralised responsibilities, while she prefers to explore whether it is really possible in the mind of Venezuelan mayors.

of government and, in turn, inefficiencies in their allocation. Water service supply, by its nature, has to be shared between different institutional bodies as the authority over water streams and basins is often shared amongst different municipalities and levels of government. Nevertheless, as Hitcher (2002) has affirmed, during the 1990s, more than one hundred regulation instruments were produced to distribute and manage responsibilities for water provisions amongst the national, state and municipal levels, producing an inevitable crisis in service organisation. Though laws for the provision of water and sanitation services (LOPSAPS, 2001 and 2007) aim to re-organise institutional responsibilities, their incomplete implementation still reduces the water sector into a series of overlapping and unclear tasks. This may be seen at two different levels. First, at the national level, where centralisation tendencies can have serious consequences in decentralising administrative responsibilities to municipalities: by 2004 only 5 out of the 15 hydrological enterprises were autonomous with respect to Hidroven (Hidroven and MARN, 2005), while just 9 out 16 enterprises were so in 2010 (Higuerey *et al.*, 2012). Second, at the more local level of the municipalities of Cristóbal Rojas and Paz Castillo, Hidrocapital has not even begun to decentralise its water responsibilities to municipalities.

Lack of water service decentralisation and of a clear understanding about water service responsibilities is clear in these two local governments. In the Paz Castillo municipality for instance, the director of the "Urbanism and Infrastructure" department affirmed that he was completely uninformed about the kind of organisational changes that the recent law would bring to the municipality; whilst the person in charge of the "Public Services" sector asserted that the only current responsibility upheld by the municipality was the management of the two public water tank trucks, given that the responsibility for the pipelined network belonged to Hidrocapital. Conversely, the mayor was aware of the on-going national decentralisation policy and of the possibility for neighbouring municipalities to associate themselves in the management and organisation of the service (*mancomunidad de municipios*), although he asserted "the municipality is absolutely not ready to assume the responsibilities for the transfer of public services … because we do not have the technical and financial capabilities to organise, manage and maintain the service, not only the distribution to the consumer but also the collection from the basin"[5].

[5] In particular, the mayor expressed his concerns over the possibility of the municipality assuming responsibility for public service delivery, especially with respect to the last FIDES reform, which reduced allocations for local governments from 40% to 28% – as well as for state governments – in order to provide funds to Communal Councils.

Box 5.1. Two main sources of intergovernmental transfers to Venezuelan municipalities

1. ***SITUADO CONSTITUCIONAL and SITUADO MUNICIPAL***: According to the Venezuelan Constitution (1999: art. 167) and the LOPPM (2005: art.141), the "*Situado constitutional*" is a budget item entrusted by the National Treasury to the states and the Capital District in each fiscal year and is equivalent to a maximum of 20% of ordinary total national revenues. Each municipality must receive not lower than 20% of this budget item from the corresponding state. Additionally, each municipality must receive the "*Situado municipal*", which is not lower than 20% of the other state revenues (outside the "*Situado constitutional*"). The LOPPM (2005: art. 141) specifies the distribution's criteria of the *Situado* as follows: 30% in equal parts for each municipality; 50% in proportion to the municipality's population; 20% in proportion to the territorial extension.
2. ***FONDO DE COMPENSATION INTERTERRITORIAL (FCI)***: Established in 2010 and officially instituted in 2011. The fund is financed from two sources (Rios *et al.*, 2012):
   - *FONDO INTERGUBERNAMENTAL PARA LA DESCENTRALIZACION (FIDES)*: The intergovernmental fund for decentralisation has been operating since 1993 to "promote decentralisation, inter-territorial solidarity and the development of the States, the Municipalities, the Capital District of Caracas, the Alto Apure District and the Community Councils, together with the support citizen participation" (*Asamblea Nacional de la Republica Bolivariana de Venezuela*, 2006: art. 1). This fund can be obtained by states, municipalities and communal councils, to finance specific programmes and projects related to the operation of institutions and the provisions of public services.
   - *LEY DE ASIGNACIONES ECONOMICAS ESPECIALES PARA LOS ESTADOS Y EL DISTRICTO METROPOLITANO DE CARACAS DERIVADAS DE MINAS E HIDROCARBUROS (LAEE)*: Through this law (established in 2000), a percentage of the national revenues from mining and hydrocarbons are redistributed amongst states, municipalities and organised communities, in order to compensate environmental damages generated by economic activities and sustain the development and improvement of life conditions for the population in the whole states and municipalities[2].

   FCI distribution was decided by the Federal Government Council in 2011 as follow: 37% for states, 28% for municipalities and 35% for communal councils (*consejos comunales*).

In the Cristóbal Rojas municipality, there was a somewhat different consciousness of – and reaction to – the forthcoming water service decentralisation. At the time of investigation, the director of the "Urban Conservation and Maintenance" department was in fact collecting a large amount of information with regards to private urban waste collection facilities – especially trucks – in preparation for the decentralisation of this service. He was aware about the approaching decentralisation process for the water sector and was ready to start collecting information for this service too. In the same way, the director of the "Urban and Rural

Planning" department was similarly conscious of the legal prescriptions but also that the consolidated situation was hardly changeable: "from a legal point of view, water and power are a responsibility of the municipality, but it has always been delegated to the national enterprises – respectively the INOS and the CADAFE (*Compañía Anónima de Administración y Fomento Eléctrico*). The municipal responsibility has always been delegated ... Hidrocapital is the institution currently responsible at the operational level, while the municipality is the responsible as a manager (*gestor*)".

Where the possibility to organise the provision of water through the *mancomunidad* is known by key municipal servants[6], this hypothesis is generally strongly supported because "in this way decentralisation is not assumed by a single municipality, but together, by the 6 Tuy municipalities and by the metropolitan area of Caracas". Joint collaboration among the Tuy Valleys municipalities is identified for instance in the balancing out of service fees: "they [Hidrocapital] are studying a way through which the areas that are paying can subsidise the areas that do not have the same capacity to pay; Hidrocapital has been studying this since last year [2005]"[7]. Nevertheless, the *mancomunidad de municipios* established in 2002 through the FONDEI (*Fondo para el Desarrollo Integral de los Valles del Tuy* – Fund for the Integrated Development of the Tuy Valleys) had not yet been involved in the water service decentralisation process. Instead it was dealing mainly with the elaboration of the Integrated Endogenous Development Plan for the Tuy Valleys, the Investment Plan in areas with regional impact, and the elaboration of other priority projects for the Investment Plan, but not water related projects.

Fiscal aspect

The two municipalities seem characterised by a general condition of non-viability of and extra-municipal dependence over financial resources. The calculation of effectiveness UGI indicators related to municipal fiscal sustainability and the intergovernmental fiscal transfers' structure reveal the presence of a highly centralised fiscal scheme. In both municipalities, local governments' incomes are deeply dependent

6 Municipal servants informed about the *mancomunidad de municipios* – as an alternative to water provision by a single municipality (LOPSAPS, 2001: art. 12) – generally became aware of it through the meetings organised by Hidrocapital with the municipalities of the Tuy Valleys in order to explain the forthcoming decentralisation policy in the water sector and organise the managerial and organisational changes (information collected through interviews with the Director of "Urban and Rural Planning" and the Mayor of Paz Castillo Municipality).

7 Information collected through interviews with the Director of "Urban and Rural Planning".

on the transfers provided by the national and state governments, when compared to the UN-HABITAT results from the second field test (UN-HABITAT, 2004a). A slightly different situation applies in the Cristóbal Rojas municipality where the financial system seems to be somewhat more viable and independent from higher level government than in the Paz Castillo municipality. This is revealed by the higher ratio of recurrent to capital budget (UGI indicator No. 2) in Cristóbal Rojas (1.25 against 0.16) and by the lower percentage of local government revenue from transfers (UGI indicator No. 3) for this municipality (44.4%) in respect to Paz Castillo (86%).

However, municipalities' fiscal reliance on intergovernmental transfers seems to be particularly deep and widespread throughout the country, as shown by the last ten years' literature concerning Venezuelan fiscal decentralisation (e.g. Mascareño, 1996, 2003; Reveron Escobar, 2003; Leal Rìos and Gotera, 2003; Gonzales Villar, 2004; Moreno, 2008; Rìos *et al.*, 2012). In the previously mentioned study concerning the opinion of Venezuelan mayors on the subject of the decentralisation process, Reveron Escobar (2003: 215-216) points out that although only 24% of the interviewed mayors declared that its municipality was unable to collect its own revenues directly, 83% of these mayors acknowledged that the major source of financial resources in their municipality depended on intergovernmental transfers and not on their own income[8]. Looking into recent figures, an IDB analysis states that, in Venezuela, "around 70 per cent of sub-national revenues come from transfers from the central government" (Rios *et al.*, 2012: 2)[9]. The fiscal dependency of local government incomes on higher level governments is also maintained by a very clear and steady transfer allocation scheme – defined by the current Venezuelan laws concerning municipal incomes[10] – which allows municipalities to easily and confidently predict the (generous) amount of funding they will receive. Box 5.1. describes

8 Venezuelan local government revenues mainly come from intergovernmental transfers (Municipal and Constitucional *Situado*, LAEE and FIDES), from own revenues collected at the municipal level (economic activity of industry and trade, services, fines and penalties, urban estate, …) and other resources coming from municipal participation in other programmes multilaterally funded (Leal Rios and Gotera, 2003).

9 The municipal fiscal dependency on national and state resources is not just high but seems also to fluctuate over the years: 35% of the entire municipal revenue came from transfers in 1986 (Mascareño, 2003), while about 50% of it came from transfers between 1998 and 2007 (Mascareño, 2003; Ríos *et al.*, 2012), 48% in 2008 and 2009 and 36% in 2010 (Ríos *et al.*, 2012).

10 The municipal income structure is mainly regulated by the Venezuelan Constitution of 1999 and by the recent Organic Municipal Public Power Act (Ley Organica del Poder Publico Municipal – LOPPM) of 2005.

the two main sources of intergovernmental transfers to Venezuelan municipalities: the Constitutional and Municipal *Situado*, and the Inter-territorial Compensation Fund, including the Intergovernmental Fund for Decentralisation (FIDES) and the Special Allocations for States derived from the Mining and Hydrocarbons Act (LAEE).

These transfer instruments are deemed to create an inefficient fiscal system (Mascareño, 1996, 2003; Leal Rìos and Gotera, 2003; De la Cruz, 2003; Reveron Escobar, 2003; Moreno, 2008; Rios *et al.*, 2012). Leal Rìos and Gotera (2003) argue for instance that the transfer system generates at least two main imbalances. The first is given by the aforementioned high disproportion between municipal revenues coming from intergovernmental transfers and those coming from their own revenue collection, produced by excessive turnout from the central government. As a consequence, the strong fiscal presence of the central government causes dependency – fiscally but also politically – on higher governments. This corresponds to a lack of incentives for municipalities to strengthen their own fiscal collection and take responsibility for the provision of services – helping to explain why, as previously mentioned, in the two analysed municipalities, drinking water services have not yet been decentralised. The second is represented by the discrepancy between the well-defined allocation scheme for sub-national governments and the absence of a corresponding expenditure design at the municipal level. These authors affirm that this system is quite inefficient as the transfer is not associated with the cost of the services while, in the case of concurrent responsibilities amongst different levels of institutions, more expenses are associated with the services' management. An additional weakness of this fiscal system is underlined by Rios *et al.* (2012) who show how this municipality revenue dependence may increase the volatility of revenues as they are affected by large fluctuations due to their reliance on oil revenues.

Institutional efficiency in delivering public services and responding to civil society concerns and welfare

In the case of the municipalities under study, at an operational level, the incomplete decentralisation process still allows Hidrocapital responsibility for the provision of drinking water. It is interesting to observe that although Hidrocapital officials are well aware that water services could be contracted out to other, non-state owned providers (according to the LOPSAPS, 2005, art: 46), they also realise that municipalities are unable to assume the service's organisation and management, and that public officials – if obliged – prefer to entrust it to the current water companies through the *mancomunidad de municipios*. For this reason, any consideration regarding the formal and institutional mechanisms in place to guarantee effective water service delivery – as identified by effectiveness UGI indicators – needs to also take Hidrocapital into account, rather than just looking to the local government.

It emerges thus that, in both the selected municipalities, mechanisms for the effective delivery of public services, such as a published performance delivery standard for water supply (UGI indicators No. 6), and for the integration of civil concerns in improving service delivery, such as a customer satisfaction survey concerning water services (UGI indicators No. 7), are not present[11]. However, somewhat different results can be found if these UGI indicators are applied to the state-owned water company. Hidrocapital has, in fact, a standard related to water "quality" and a specific office charged with the definition and control of this standard at their headquarters in Caracas and Yare. This standard is established at the state level because Hidrocapital covers the Federal District, the Vargas State and the Miranda State[12]. On the contrary, neither the municipalities nor Hidrocapital have organised a customer satisfaction survey, although the water company provides local offices where people can complain about the service, and a hotline to report grievances[13].

Similar considerations can be made about the presence of a vision statement with clear objectives and targets for municipal development (UGI indicator No. 8) and more specifically for the provision of water services. Both municipalities are part of the project *Alcaldia Digital*, endorsed by the Ministry of Science and Technology with the objective of "building the technological and managerial conditions that can allow municipalities to effectively process knowledge and information, in order to optimize local management quality and promote interactions with citizens". This project asks municipalities to provide a vision statement concerning their future developments: only Cristóbal Rojas had one, affirming their intent to guarantee all citizens "the satisfaction of their requests about goods and services in the municipal areas of competence", but without any explicit reference to water services[14].

---

11 Information collected through interviews with municipal public servants in the Paz Castillo and Cristóbal Rojas municipalities.

12 Information collected through an interview with the Community Coordinator of the Losada Ocumarito System.

13 The hotline number is 0800-7682253: 0800-potable.

14 The website of *Alcaldia Digital* (http://www.alcaldias.gov.ve/portal-alcaldias/portal.html) seems no longer to be operational, according to numerous unsuccessful attempt to access it at the end of 2013 and during 2014. The Cristóbal Rojas municipality website actually provides the same mission statement (http://www.charallave.gov.ve, last visit: 6th of September 2014). In the case of Paz Castillo, their vision was not yet available on the *Alcaldia Digital* project website (last visit: 26th of May 2007), while

Looking to the water company, on the contrary, while no official vision of the company is available – at least on the website – Hidrocapital administrators clearly put forward the company's aim to guarantee a good quality service for all Venezuelans. The community coordinator of the Losada Ocumarito system affirmed that the new Hidrocapital vision was access to drinking water for all people, based on the assumption that human beings have the right to water. The General Manager of the Losada Ocumarito system stated that Hidrocapital had the objective of providing water 365 days per year, 24 hours per day and to all people in its territorial jurisdiction.

5.2. Equity in water provision policies with disparate results in water access

Equity in local governance is investigated using two main sub-criteria: distributional equity, which focuses "on the policies, processes, tools or mechanisms present for access to basic services" (including water services, in the specific context of this analysis), and procedural equity, which represents the "equity in decision-making under the principle of participatory governance" (UN-HABITAT, 2004a: 21). Results from the calculation of UGI indicators (see table 5.2.1.) related to the distributional sub-criterion show that, despite the only recent approval of a specific statement acknowledging citizens' right to access to water services (UGI indicator No. 9) – which was published after the time of the field investigation – the other distributional equity indicators perform well. In fact, first, the percentage of households with water connections is quite high in the two selected municipalities (UGI indicator No. 12A) and, second, there exists a pricing policy at the national level which takes into account the needs of poor households, providing lower rates compared to the prices applied to the industrial and commercial sectors (UGI indicators No. 12B and 12C)[15]. Actually, the previous (1993) and current (2011) tariff structure along with the normative framework which belong to the *Ley Organica para la Prestacion de los Servicios de Agua Potable y Saneamiento* constitute a legal basis on the one hand to ensure an efficient and cost-effective service together with affordable water

---

the actual municipality website (http://www.pazcastillo.gob.ve) provides a mission statement that aims to promote the "common good" but without any reference to urban public services (last visit: 6th of September 2014).

15 In the "UGI Conceptual Foundation and Field Test Report" (UN-HABITAT, 2004a: 36) it is suggested that the indicator on "pro-poor pricing policies" can be disaggregated as follow: A. Proportion of households with direct access to water (piped connection). B. Is there a water pricing policy (yes/no). C. Is the metered price of water per litre lower for households below the poverty line (yes/no)?

prices for low-income residents and, on the other, to try to build a new culture in relation to the payment of water services.

For the new LOPSAPS to be fully implemented, responsibility for the organisation and management of water service provisions needs to be shared amongst all of the actors involved in the service delivery process. At the moment, however, the capacity and the willingness of these stakeholders does not appear to be easily taken for granted (see for instance the consideration previously presented on municipal decentralisation). At the same time, the "literal" application of this new law for water and sanitation services could also lead to a possible prioritisation of market mechanisms over the equitable access to drinking water (as in the case of a private management and operation of service provision). Besides the unquestionable importance of the new LOPSAPS in stating a clear framework to ensure equitable access to water, it appears very difficult – if not impossible at this time – to assess to what extent this law could effectively guarantee equitable access to water in the future and to anticipate any possible market-oriented scenarios opened by this.

From the perspective of procedural equity, as per the UGI indicators, women are poorly represented in municipal councils (UGI indicators No. 10 and No. 11) and incentives exist to informally arrange improved water access (UGI indicator No. 13, also concerning distributional equity), which in any case does not guarantee *per se* low income communities equity in accessing water or in the decision-making process, as communities have first to establish an MTA. However, the bulk of considerations associated with procedural equity are above all described in the following section of this chapter related to "participation", as the procedures and mechanisms entailing the inclusion of all stakeholders involved in the water delivery process, including the decision-making phase, are widely described in relation to participative democracy (§ 5.3.2.).

In sum, it seems possible to affirm that, despite the possible overestimation of the percentage of households connected to sustainable piped water provisions, if we also consider water quality and service frequency, the current normative framework established at the national level seems to give priority to an equitable and inclusive access to drinking water. This is done through a water tariff structure and subsidies – amongst other factors – aimed at improving access to drinking water for low-income communities in particular. These conclusions are based on the elaboration of two profiles (distributional and procedural equity) presented without consideration to any differences between the two municipalities. This is because the majority of the equity UGI indicator results have been worked out on the basis of regional or national data, which concurrently apply to both municipalities. This also allows us to partly generalise our findings to cover the entire country.

Table 5.2.1. UGI equity indicators and equity sub-criteria

EQUITY UGI INDICATORS				*EQUITY SUB-CRITERIA*	
No.	**Indicator**	**Data (X)**		***Distributional equity***	***Procedural equity***
		Paz Castillo	**Cristóbal Rojas**		
9	Citizens charter for water services (CCS)	No Yes (2008)	No Yes (2008)	•	○
10	Percentage of women councillors (WC)	28.58%	28.58%	○	•
11	Percentage of women in key positions (WK)	*n. a.*	*n. a.*	○	•
12A	Percentage of households with water connection (HH wat)	74.82%	92.21%	•	○
12B	Existence of pro-poor policy (PPC)	Yes	Yes	•	○
12C	Is water price cheaper for poor settlements? (WP)	Yes	Yes	•	○
13	Incentives for informal arrangements to improve water access (IA)				
	a. Informal provisions allowed	Yes	Yes	•	•
	b. Informal provisions with restrictions	Yes	Yes	•	•
	c. support for informal provisions	Yes	Yes	•	•
	Equity sub-index	**0.63**	**0.67**		

• = indicator considered belonging to the sub-criterion
○ = indicator considered not belonging to the sub-criterion

Policies, processes, tolls and mechanisms for access to basic services (distributional equity)

Right to water[16]

Venezuelan laws make frequent reference to the public character of water resources and the need for their equitable access[17]. The 1999

[16] According to the UN-HABITAT definition (2003c: 12), a citizens' charter for basic services is directly linked to their right of access to these services: "presence or absence of a signed, published statement (charter) from the local authority which acknowledges citizens' right of access to basic services".

[17] Venezuela ratified the International Covenant on Economic, Social and Cultural Rights on the 10th of August 1978 and – as all the states ratifying the covenant – has the obligation "in relation to the right to water, such as the guarantee that the right will be exercised without discrimination of any kind (art. 2, paragraph. 2) and the obligation to take steps (art. 2, para. 1) towards the full realization of articles 11, paragraph 1, and 12. Such steps must be deliberate, concrete and targeted towards the full realization of the right to water" (Committee on Economic, Social and Cultural Rights, General Comment No. 15, 2002, art. 17).

Constitution (art. 304) states "all waters are goods in the public domain belonging to the Nation and indispensable for life and development. The law shall prescribe the provisions required to ensure its protection, utilisation and recovery, with respect to the phases of the hydrological cycle and the legal criteria for land use and regulation". On the basis of this assumption, the National Plan for the Economic and Social Development of the Nation 2001-2007 (*Plan Nacional de Desarrollo Economico y Social de la Nacion*) establishes that in order "to guarantee the universal and equitable enjoyment of social rights", it is necessary to ensure the provision of a safe, secure and ecologically sustainable environment and housing, which includes "drinking water and sanitation services", while achieving the same performance in service quality, coverage and sustainability (RBV, 2001, paragraph. 2.1.4.2). In the same way, the LOPSAPS (2001, art. 3.b.) identifies "the access of all citizens to drinking water and sanitation service provisions" as one of the principles underpinning delivery of the public services regulated by this law.

Box. 5.2. Ley De Aguas (Gazeta Oficial 02/01/07)

Artículo 5. Los principios que rigen la gestión integral de las aguas se enmarcan en el reconocimiento y ratificación de la soberanía plena que ejerce la República sobre las aguas y son:

1. El acceso al agua es un derecho humano fundamental.
2. El agua es insustituible para la vida, el bienestar humano, el desarrollo social y económico, constituyendo un recurso fundamental para la erradicación de la pobreza y debe ser manejada respetando la unidad del ciclo hidrológico.
3. El agua es un bien social. El Estado garantizará el acceso al agua a todas las comunidades urbanas, rurales e indígenas, según sus requerimientos.
4. La gestión integral del agua tiene como unidad territorial básica la cuenca hidrográfica.
5. La gestión integral del agua debe efectuarse en forma participativa.
6. El uso y aprovechamiento de las aguas debe ser eficiente, equitativo, óptimo y sostenible.
7. Los usuarios o usuarias de las aguas contribuirán solidariamente con la conservación de la cuenca, para garantizar en el tiempo la cantidad y calidad de las aguas.
8. Es una obligación fundamental del Estado, con la activa participación de la sociedad, garantizar la conservación de las fuentes de aguas, tanto superficiales como subterráneas.
9. En garantía de la soberanía y la seguridad nacional no podrá otorgarse el aprovechamiento del agua en ningún momento ni lugar, en cualquiera de sus fuentes, a empresas extranjeras que no tengan domicilio legal en el país.
10. Las aguas por ser bienes del dominio público no podrán formar parte del dominio privado de ninguna persona natural o jurídica.
11. La conservación del agua, en cualquiera de sus fuentes y estados físicos, prevalecerá sobre cualquier otro interés de carácter económico o social.
12. Las aguas, por ser parte del patrimonio natural y soberanía de los pueblos, representan un instrumento para la paz entre las naciones.

Recently, Venezuela has also explicitly stated the right of access to water services (UGI indicator No. 9) through the "Water law" (*Ley de aguas* – published on the 2nd of January 2007) which states that "waters, being a public domain, cannot be part of the private domain or of any natural or juridical person" (art. 5.10) and that "the access to water is a fundamental human right" (art. 5.1) (see box 5.2.). The current National Plan for the Economic and Social Development of the Nation (2013-2019, b.II.) confirms that the country has "to continue supporting the acknowledgment of the access to drinking water as human right in all the national and international contexts" and that it is necessary "to universalise drinking water services and its availability to all people, guaranteeing equity" (*Politica y Programas del Sector Ambiente*, 47). This National Plan also considers that the country has to ensure access to water as one of people's basic needs and to "strengthen and improve the systems of drinking water throughout the national territory, ensuring and improving the continuity of drinking water services up to 250 litres per person, through the construction of a hundred aqueducts at the national level, to guarantee the production of 4,000 billion cubic metres of drinking water" (art. 3.4.1.3.). At the local government level – according to the interviews held with public servants in the Paz Castillo and Cristóbal Rojas municipalities – no further legislation has been created with respect to the right to water, making reference only to national legislation.

Pro-poor access to drinking water

According to UN-HABITAT (2004a: 36), an equitable distribution of basic services – and drinking water in particular as a key service – is a complex issue which can be measured and, above all, understood through a combination of information concerning "equitable access to water, subsidised water, water consumption slabs and comparison of residential water price to commercial or industrial sectors". The situation stemming from the calculation of equity UGI indicators in the two selected municipalities – but, again, also in the wider Caracas Metropolitan Region – appears characterised by a high proportion of urban households with direct access to piped water connections (UGI indicator 12A), and by the presence of a pro-poor pricing policy for water (UGI indicator 12B) and lower water tariffs for households below the poverty line (UGI indicator 12C).

Direct access to water (piped connections)

At the national level, Venezuela seems to have realised great improvements in water service provisions, given the fulfilment of the Millennium Development Goal target concerning access to water 10 years ahead of schedule (INDIA, 2006; Lacabana *et al.*, 2008), through a modest but progressive increase in urban drinking water provision coverage between 1998 to 2004 (Hidroven and MARN, 2005). According to the last national population and settlement census (2011), access to piped water has further improved at the national level (JMP, 2014). Public utility coverage through piped water connections in the two selected municipalities is relatively high: the percentage of households with piped access to water, with respect to the total accessing improved water sources, is 82.06 in Paz Castillo and 91.24% in Cristóbal Rojas (INE, 2011).

Table 5.2.2. Frequency of piped water (percentage)

	Every day	Every 2 or 3 days	Once per week (every 8 days)	Once every 15 days
Cristóbal Rojas (municipality)	26.29	39.96	24.08	9.67
Charallave (Parish)	29.53	44.39	19.31	6.77
Paz Castillo (municipality)	25.89	49.79	14.40	9.92
Santa Lucia* (Parish)	25.89	49.79	14.40	9.92

Source: INE, 2011, *National Census on Population and Settlements.*
*Data provided by the national census are equal for the municipality of Paz Castillo and the parish of Santa Lucia.

Nevertheless, when considering piped access to water, the frequency and quality of the service must also be taken into account. Concerning frequency, for example, Hidrocapital (2005) estimated that water was rarely available all day long, reaching just 12 hours per day in both Santa Lucia (Paz Castillo municipality, where the community of the Hacienda el Carmen is located) and in Charallave (Cristóbal Rojas municipality, where the community of Paso Real 2000 is located). The 2011 national census provides more precise figures on the frequency of piped drinking water (see table 5.2.2.) showing that the service is currently available every day for just a quarter of the population. Due to the discontinuous provision of drinking water in the Tuy Valleys, households belonging to low-income communities are obliged to rely on different water supply systems, such as water trucks[18], or, at best, to collect water in drums for when it is unavailable. As a consequence, the quality of water in the

[18] This practice occurs in the neighbouring areas of the analysed communities. For example, many of the new squatter settlements situated next to the *Hacienda el Carmen* (e.g. *Limon, Los Munos, 23 de Enero*) rely on distribution by municipal or private water trucks (together with the illegal connections built over the pipelines and realised by the community and the *mesa tecnica de agua* in the *Hacienda el Carmen*).

main piped network can be seriously affected by residents' careless water storage and by the bacteria that might develop in the pipes because of intermittent frequency, such as during planned water rationing[19], or when the water pressure is lowered due to new illegal connections.

Pro-poor pricing

In the period between the late 1960s and early 1970s – the INOS (*Instituto Nacional de Obras Sanitarias*) did not support a policy of cost recovery for water services due to the administrators' political vested interests and the possibility of not charging for piped water because of the previous huge oil incomes (Francisco, 2005; Wilpert, 2005). Public hydrological enterprises belonging to Hidroven have often continued not to charge for water provisions – especially from the low-income communities – with the result that, at the beginning of 2004, 65.73% of water consumption was not charged for amongst Hidroven subsidiary companies (Hidroven and MARN, 2005: 35). This percentage reached 71.91% in the areas covered by Hidrocapital (2005) (Capital District, Vargas State and Miranda State). The reason seemed to be based on "the high cost for water companies of retrofitting residential connections and installing micrometers" (IDB, 2011: 3), but also on the political choice to exclude those with low financial resources and those who benefited from the service only during the night or once a week from payment (Francisco, 2005; INDIA, 2006).

Today, the approach to the financial dimension of water supply tends to privilege payment for water by all users (included low-income dwellers) but according to a pro-poor pricing allowing them to afford the service. Pro-poor pricing is based on both tariffs and subsidies. So, first, although the new LOPSAPS does not directly establish specific tariffs for water services[20], it defines a new approach for their calculation, which is supposed to unite equitable access to water services with economic efficiency (Baez, 2001). Tariffs shall include all the administrative, operative and maintenance costs together with the costs concerning investment in rehabilitation, expansion and capital remuneration (art. 85). However, the IDB (Rios *et al.*, 2011: 1) notes how "as a consequence of the government's social policies, the current tariff structure does not include investment costs or recovery of capital".

19 As we will explain in the next chapter, to cope with this shortcoming, the frequency of piped water provisions is planned (*racionamiento programado*), allowing residents to organise themselves in order to store and use water (Cariola and Lacabana, 2005b).

20 Tariffs shall in theory be defined by the National Superintendency of Drinking Water and Sanitation Services and submitted to the approval of municipal authorities (LOPSAPS, 2005: art. 83).

Tariffs shall also be composed by a fixed and a variable charge – where the latter reflects the effective quantity of water consumption (art. 86), while envisaging subscribers' categories – defined by the National Superintendency – who shall be charged with different tariffs according to the "economic activities and the socio-economic conditions belonging to the residential subscribers" (art. 87). Currently, the *Gaceta Oficial* No. 39788, published on the 28th October 2011 provides the methodology, formulas, models and technical criteria to regulate tariffs for water and sanitation services provided by the regional hydrological enterprises affiliated to HIDROVEN, and substitutes the previous *Régimen Tarifario para la prestacion de los Servicios de Acuaductos y de Recoleccion, Tratamiento y Disposicion de Aguas Residuales* (*Gaceta Oficial* No. 35161 of the 1st March 1993). This law regulates and determines the reference medium price for water per m³, according to which tariffs for social, residential, commercial and industrial building usage is calculated[21].

The social tariff is applied exclusively to residential buildings located in geographic areas where a wide part of the low-income population lives; this tariff is applicable to all of the six hydrological systems and on up to 40 m³/month consumption only. For Hidrocapital, in 2006 – at the time of field data collection – the social tariff was 242.78 Bs/m³ (= 0.24278 Bs.F), equivalent to 0.039 US dollars[22]; from 2011, the social tariff has been calculated according to a fix and variable component: the fix component is up to 15 m³/month and is equivalent to the reference medium price (1.55 Bs.F = 1,550 Bs) multiplied for the use factor (social use = 0.5) and again for 0.5, which means 0.3875 Bs.F per m³, equivalent to 0.062 US dollars. The variable component is between 15 and 40 m³/month and it is equivalent to the reference medium price multiplied by the use factor and again for 0.75, which means 0.58125 Bs.F per m³, equivalent to 0.092 US dollars (*Gaceta Oficial* No. 39788, published on the 28th October 2011). In reality, low-income dwellers receiving water quite regularly and charged with the social tariff, do not generally pay this price, as a consequence of the previous non-cost recovery policy and according to a general conception of public services gratuity (Cariola and Lacabana, 2005b; Francisco, 2005).

21 With the previous tariff rules, each water system developed its own tariff structure. Tariffs were thus identified by Hidrocapital through six tariff types (according to the localisation of the system where water is collected) and 5 kinds of different uses (residential, social-residential, commercial and 2 types of industrial uses). For instance, tariff 4 was applied in the Losada-Ocumarito System – together with the Barlovento System. The six different tariffs were established according to the operation costs connected with the water and sanitation infrastructure and the hydrological capacity.

22 www.hidrocapital.com.ve.

Just after the LOPSAPS publication, Baez (2001) underlined how the tariff policy outlined in the new law is characterised by a service delivery which has to be both efficient and equitable, where rights and responsibilities are shared amongst the suppliers and the beneficiaries: the supplier is to deliver an efficient service, the subscriber is to pay for it, while tariffs are to be equitable and able to guarantee the economic effectiveness of water services. In this perspective, as the payment of bills is compulsory, the hydrological enterprises are making improvements in connecting water meters at the household or condominium level. Hidroven and MARN reported in 2005 that, with funds from the Inter-American Development Bank (IDB), three hydrological enterprises (Hidropaez, Hidrollanos and Hidrocaribe) have already set up 75 thousand meters. But, recent figures from the IDB (Rios *et al.*, 2011: 3) reveal that despite some progress in the sector[23], "... there is still a high level of unbilled water (56%), due mainly to the number of people who do not pay for service and the low level of metered usage".

Second, equitable access to water services is also ensured by the subsidies' framework foreseen by the LOPSAPS. Subsidies are constituted in order to "finance minimum consumption necessary for an average household" and "focus on the households with scarce resources" (art. 95). They can be "demand-based", which is a financial aid to the subscriber, "offer-based" meaning a financial support to investments, or "cross-cutting subsidies" envisaging a higher tariff to partly cover the payment of low-income households (art. 96-101), while either the national executive power, the states, the municipalities or the metropolitan districts shall be charged with their application.

The established tariffs and subsidies framework for water and sanitation services is worked out according to the institutional intent to make people pay for the service: water service payment is proposed to low-income communities and collected amongst its members by the *mesas tecnicas de agua.* According to Francisco (2005), MTAs have dramatically improved payment collection levels, as they have been able to communicate that the hydrological enterprises were not charging for the water but for the service itself. In this perspective, service users access to and benefit from a common resource for free, whilst paying for the treatments to make water drinkable (electricity to pipe water, chemical substances to purify it, pipeline maintenance costs, workers' costs, etc.) and have, at the same time, the option to complain about the service. The ex-community coordinator of the Losada Ocumarito system affirmed that:

23 "Between 1998 and 2008, the volume of metered water billed, as a percentage of total volume produced, increased from 20% to 36%; at the same time the meter-reading rate increased from 14% to 17.5% (Rios *et al.*, 2011: 3).

> it was to create the culture to pay for water … to create the condition of co-responsibility as stated in the constitution, to assume the responsibilities; it was as to create, in a certain way, inside the dream, to wake it up and to build a society with responsibility. It used to be though that the responsibility of the service belonged just to the state … We believe in the process of power through the participation of common people to solve the problem, as in the case of the MTA.

In conclusion, it seems possible to argue that Venezuelan water policies and strategies are pro-poor. With respect to the theoretical framework proposed to investigate distributional equity (Crane and Daniere, 1996), the current framework for tariffs and subsidies – also based on the MTA's contribution to informing about the service and collecting payments – seems to address the supply/demand and individual/community dichotomies in access to the service. Thus, on the one hand, water access costs for low-income residents (reflecting supply factors) are dramatically reduced through social tariffs and subsidies while, on the other, the water access values carried by low-income residents (reflecting demand factors) have been met through the provision of an efficient and reliable service and the right to claim for maintenance and obtain repairs. At the same time, MTA involvement in the implementation of this process allows the whole community, and not just a few individuals, to access the service.

Nevertheless, there is at least one critical aspect of this normative framework, which could undermine equitable access to water for low-income communities: private sector participation in water service provision and its consequences on an equitable distribution of water services. Once the decentralisation process is accomplished, municipalities will be free to provide the service directly or to open up to international bidders, which could include public as well as private competitors (LOPSAPS, 2001: art. 46). Two different positions have been expressed with respect to the possible participation of the private sector with the resultant consequences.

An optimistic scenario is drawn by Cristóbal Francisco[24] (2005). He recognised in effect that the new LOPSAPS could allow private sector participation in the water sector – also from transnational enterprises – from a normative point of view, with the risk of raised tariffs and non-existent subsidies. Nevertheless, he believes that this situation will not materialise, and this law will therefore not undermine equitable access to water services, for two main reasons. The first rests on national policy principles – regulated by the last Constitution (1999) – which considers water as a public/common good, forbids monopolies in the

---

24 Cristóbal Francisco was the Hidroven president and vice-minister for water, and current Minister of Environment (*Ministro del Poder Popular para el Ambiente*).

hydrological sector and allows concessions only for a limited time. Accordingly, Francisco argues that private participation has been added to the LOPSAPS in order to allow cooperatives to participate in the water service provision. The second relates to the option for municipalities (or *mancomunidad de municipios*) to decide not to open up to international bidders and instead continue to hand over the service to the public sector, through the same or other public water companies. Thus, the author affirms "the task is to create a level of consciousness among people which could prevent mayors and municipal councils from handing the service to transnational enterprises" (Francisco, 2005: 151).

Conversely, a more pessimistic scenario was foreseen by Santiago Arconada (2005: 196), who regarded the new LOPSAPS as a "serious neo-liberal error" insofar as it misunderstood the rationale behind community work. Thus, he underlined how the LOPSAPS "is deeply pervaded by a market halo which has nothing to do with the management of water and sanitation as a place belonging to people's power and strategies, which cannot result in the transformation of service into an economic good" (Arconada, 2005: 197). In this scenario, equitable access to water could, in effect, be undermined by profit-oriented strategies.

Equity in decision making – participatory governance (procedural equity)

Incentives for the development of formal/informal systems and arrangements to access water services

As previously discussed, the current Venezuelan governance system to improve water service provisions – especially for low-income residents – is based on formal and informal systems of accessing drinking water, as well as formal and informal arrangements in the service delivery process. In the two selected municipalities – as in most Venezuelan municipalities – formal practices are represented by piped networks, public or private household distribution through licensed tankers, wells or bore-wells; while informal practices include illegal connections to the main public pipes or private distribution through non-licensed tankers[25]. Insofar as Hidrocapital intervenes in assisting low-income communities to build their piped network and in "regularising" the previous illegal water provision network, it is possible to argue that there is an explicit effort to improve the organisation of and to further develop these kinds of connections without removing them. From this perspective, the new LOPSAPS

[25] As pointed out by Cariola and Lacabana (2005b), "informal" connections to the public piped network can be defined as "illegal" practices that have been permitted and sometimes supported by Hidrocapital.

could be regarded as a normative framework aimed at normalising and institutionalising informal practices and thus as an incentive for informal arrangements to provide water services (UGI indicator No. 13), based on the previous illegal practices but regulated through a new normative tool. This incentive, however, involves the establishment of an MTA to contribute to the governance process.

Besides, informal, local and community practices could face numerous limitations when participating in local governance arrangements – frequently characterised by the absence of codified rules – given that the better networked and organised groups have a much higher chance of being listened to than powerless voices or individuals outside the group (see Maloutas and Malouta, 2004 and chapter 1). As a result, communities that face difficulties and internal tensions when organising themselves to lobby the municipality and the water company will be disadvantaged in accessing water service provisions and in the decision-making process. As a matter of fact, in both municipalities, Hidrocapital officials and municipal servants affirm that community representatives who are well organised and who constantly complain about the service are much more likely to make their voices heard and obtain the service rather than scattered communities members experiencing the same water shortages.

Procedural equity

Within the equity indicators of the UN-HABITAT Urban Governance Index (see table 5.2.1.), procedural equity is largely left to the issue of gender balance. In the case of the two chosen municipalities, data shows that 2 out of 7 councillors were women in the 2005 municipal elections in both cases (UGI indicator No. 10), while data is not available concerning the percentage of women in key positions in the local governments (UGI indicators No. 10 and No. 11). However, when speaking about procedural equity related to improved water service delivery in Venezuela, the gender perspective is not enough as there are at least two other fundamental categories of local actors deserving of specific attention alongside women. These include the Hidrocapital technicians in their relationship with community residents to realise the formal water network, and the role of children in the government strategy to make people aware of the value of water and ways to preserve the resource.

So, in the former, Cariola and Lacabana (2005a, 2005b) stress, for example, to what extent the current form of "participatory governance" is based on the collaboration between community members and Hidrocapital technicians: this cooperation allows engineers working for the hydrological enterprise to understand how the community operates in order to propose efficient solutions, while simultaneously also allowing residents to improve the preservation of water, understand how it is produced and how the service works. Decisions concerning the improvement of the water service are thus taken in a shared manner, relying on both community and technical knowledge. In the latter, children are an essential actor involved in the information process concerning the correct usage of drinking water through the *Mesas Tecnicas Escolares de Agua* (MTEA). These are small "organisations", composed of children, teachers, parents, MTA and municipal members, with the aim of teaching young students the importance of water as a "vital fluid" and how to respect water and not waste it. As a result, children belonging to the MTEA are expected to contribute to raising awareness among family members and neighbours of possible water issue solutions and good practices.

Nevertheless, beyond the incentives for informal arrangements and the gender perspective, the analysis concerning equity in decision-making is mainly related to the processes and mechanisms dealing with participative democracy included in the following section.

5.3. Democracy under tension ...

Representative democracy and participative democracy at the local level are the two sub-criteria adopted to assess participation in governance, as per the UN-HABITAT UGI framework. Calculation of UGI indicators is complete for the representative democracy sub-criterion while just one indicator (UGI indicator No. 17: People's Forums) out of the two concerning participative democracy is available, so that for the latter sub-criterion, the creation of a narrative profile mainly relies on recent critical literature concerning citizen participation in decision-making processes at the local level, in Venezuela. Amongst the participation indicators for which data is available, only the indicator related to voter turnout scores quite low whilst the other two are good as they reveal the existence of mechanisms and procedures for participation in local governments, as defined by UN-HABITAT (UGI indicators No. 14 and No. 15). Differences between the two studied municipalities of Paz Castillo and Cristóbal Rojas with respect to the participation sub-index are negligible (sub-indexes are respectively 0.73 and 0.75) and depend only on a minor variation concerning municipal voter turnout (UGI indicator No. 16).

Narrative profiles provide thought-provoking insights into the state of participation in local governance in Venezuela. First, in both municipalities, representative democracy – as defined in accordance with UN-HABITAT indicators – presents two main characteristics: on one hand, voter turnout in municipal elections is very low, even though the possibility exists to directly elect mayors and councillors; on the other, abstention rates in national elections are quite limited, although a very low voter turnout in municipal elections occurs. Figures on elections at the national level show that these considerations might be generalised, with minor discrepancies, to all of the Venezuelan municipalities (see table 5.3.2.). For considerations on electoral democracy, and more precisely on the possible shortcomings of free and fair elections, refer to the work of other authors such as Brewer-Carías (2010) and Penfold (2010), for instance.

Second, concerning participative democracy, despite an active and abundant participation of organisations working for improving water provisions through a variety of associational forms (*Mesas Tecnicas de Agua*, *Consejos Comunitarios de Agua*, *Cooperatives*, etc.), the way in which citizen participate in water supply upgrading highlights some significant limitations concerning urban governance at the municipal level. Actually, the municipality appears to be largely excluded from governance agreements and relationships for improving access to drinking water provisions as, on one side, these arrangements strongly rely on the information shared and decisions made between the MTAs and the water company, from which the municipality is almost systematically excluded and, on the other side, because of the absence of regular and formal meetings amongst the actors involved (and thus the local government as well) in the service delivery process.

Third, investigating the specific features and flaws of participation in the water sector confirms more general concerns about the relationship and mutual influence between representative and participative democracy in Venezuela. Despite most scholars recognising the new opening up towards participative democracy, some stress the positive aspects and potentials of strong citizen participation over broader democracy, while others point out the damaging effects that the current government has had on – especially representative – democracy. For the first group, Chavez's government has been able to put into practice a "participatory" (Wilpert, 2002; Wagner, 2004a), "deepened" (Gibbs, 2006) and "inclusive" democracy (Figueroa, 2006), which has been successful in realising a two-fold objective "complement[ing] representative democracy through participative democracy" (Wagner, 2004a). According to this view, the increasing involvement of the lower strata in redistributive economic policies and social programmes – aimed at providing basic services and sharing decision-making processes – has been accompanied by a renewed confidence in democracy at the national level and the resulting massive voter turnout, mainly from the poor. The analysis conducted in the two municipalities and low-income communities illustrates how dwellers involved in the water committees' programmes with state-owned water companies have intensively contributed to enabling people to vote for President Hugo Chavez in national elections[26].

Table 5.3.1. UGI participation indicators and participation sub-criteria

PARTICIPATION UGI INDICATORS				*PARTICIPATION SUB-CRITERIA*	
No.	**Indicator**	**Data (X)**		***Representative Democracy***	***Participative Democracy***
		Paz Castillo	**Cristóbal Rojas**		
14	Elected council (EC)	Yes = 1	Yes = 1	•	○
15	Locally elected Mayor (LEM)	Yes = 1	Yes = 1	•	○
16	Voter turnout (VT) (municipal elections in 2004 and 2005)	34.04%	37.66%	•	○
17	People's forum (PC)	Yes = 1	Yes = 1	○	•
18	Civic associations per 10,000 pop (CA)	*n. a.*	*n. a.*	○	•
	Participation sub-index	**0.73**	**0.75**		

Legend:
• = indicator considered belonging to the sub-criterion
○ = indicator considered not belonging to the sub-criterion

For the second group, the structural, cultural and institutional changes introduced by Hugo Chavez in order to support citizen participation, have further undermined Venezuelan representative democracy through reproducing the "undemocratic democracy" system already established by the Punto Fijo regime (Derham, 2002), "subjugating" democracy (Penfold, 2010) or making it definitively fall (Levine, 2002), "unravelling representative democracy" (McCoy and Meyer, 2004) or "dismantling democracy" in Venezuela (Brewer-Carías, 2010). This section of critical literature (see Levine, 2002; McCoy, 2004; Gil Yepes, 2004; Mascareño, 2006; Brewer-Carías, 2011; Córdova Jaimes, 2012, for more references) shares the opinion that the Bolivarian Revolution aims at establishing a direct relationship between the president and the citizenry, that is, the state and society, and that it "creates mechanisms of direct democracy while weakening mechanisms of representative democracy" (McCoy and Myers, 2004: 3). Evidence from the case study confirms that citizen participation in improved drinking water provisions rests almost entirely

[26] Community leaders in the *Hacienda el Carmen* settlement affirmed that – during the 2004 recall referendum – they used Hidrocapital's trucks to collect people in the settlement and carry them to the ballot box. Some leaders also affirmed that they helped people go to the polls only if they wanted to vote for Hugo Chavez.

on the direct and more informal co-operation between Hidrocapital and the MTAs, which sits precisely between a state institution and a civil society organisation, excluding the intermediate level of the municipality from the governance process. In this case, participative democracy overtakes the procedures of representative democracy, clearly posing the problem of the long-term sustainability of these arrangements.

Representative democracy

UGI indicators concerning representative democracy (14, 15 and 16) emphasize poor citizen participation in local representative democracy. According to the 1989 *Ley Organica de Regimen Municipal* (Organic Municipal Government Act) and the 2005 *Ley Organica del Poder Publico Municipal* (LOPPM – Municipal Public Power Act; further reformed in 2010), municipal councils and mayors are directly elected by citizens. In other words the score is equal to one for UGI indicators No. 14 and No. 15, in both selected municipalities. Nevertheless, the possibility for citizens to elect their representative local bodies at the normative level does not correspond to a high voter turnout for both municipalities, with minor differences amongst them (UGI indicators No. 16; see also table 5.3.2.). The voter turnout in the last municipal elections (mayoral elections, 2013) in the two selected municipalities scored quite low (accounting for 58.06% in Paz Castillo and 53.69% in Cristóbal Rojas), but higher than the previous municipal elections in 2004 and 2005 (an average turnout of 34.04% in Paz Castillo and 37.66% in Cristóbal Rojas; see Annex 3 for further details).

Amongst the many factors contributing to the high percentage of abstention at the municipal level, some seem to belong to a more general trend underway at the beginning of the "democratic era" in 1958, whilst others rely much more on the Chavez government context. The reasons for limited electoral participation at the local level appear therefore to be multiple: electors' tiredness due to the number and the proximity of electoral processes, together with the closeness of presidential elections to municipal ones (Velasquez Trujillo, 2005; Castellanos Vasquez, 2005); lower charges to elect and the resulting minor interest in electing municipal representatives (Castellanos Vasquez, 2005; Reyes, 2013); distrust in the electoral system with respect to its reliability and transparency (Velasquez Trujillo, 2005; Castellanos Vasquez, 2005; Maingon, 2009); lack of a credible political offer by the candidates in the running for Mayor and/or the lack of a credible alternative to the favourite candidate (Velasquez Trujillo, 2005; Maingon, 2009).

With respect to the 2005 municipal elections, two more causes can be added. First, abstention from voting as a sign of protest, either suggested by the current opposition to Chavez's government in order to express disagreement toward the current political power (Castellanos Vasquez, 2005) or in order to protest against the fragmentation and disorganisation of the opposition (Velasquez Trujillo, 2005; Helliger, 2005; Maingon, 2009) or against the imposition of certain governors belonging to the Fifth Republic Movement (*MVR – Movimiento Quinta Republica*) but not supported by grassroots leaders (Helliger, 2005). Second, abstention from voting because of "the incomplete institutionalisation of the participatory, '*protagonic*' democracy often posed by Chavez as an alternative to representative democracy" (Helliger, 2005: 9; Maingon, 2009). Helliger (2005: 29) argues that Chavez's success in the recall referendum was largely based on a national citizen mobilisation that did not occur for the municipal elections: different voter turnouts in the municipal and national elections underlines for the author "the tensions between mobilisation politics and electoral politics". Figures for national and municipal elections in the last ten years of electoral events confirm that in the municipalities of Paz Castillo and Cristóbal Rojas abstention rates have also been much higher for local elections than for national elections and referenda (table 5.3.2.).

Table 5.3.2. Abstention rates in the Venezuelan national and municipal elections from 2004 (in percentage)

	Paz Castillo	Cristóbal Rojas	Miranda State	Venezuela
National				
Presidential elections, 2013	17.9	16.8	19.7	20.3
Presidential elections, 2012	18	16.4	19.2	19.5
Constitutional Reform Referendum, 2009	28.1	26	27.6	29.7
Constitutional Reform Referendum, 2007	*n.a.*	*n.a.*	*n.a.*	44.1*
Presidential elections, 2006	24.8	23.0	25.4	25.3
Recall Presidential Referendum, 2004	34.0	25.4	30.1	30.1
Municipal				
Municipal elections, 2013 (Mayors)	46.3	41.9	n.a.	41
Regional elections, 2008 (Mayors)	38.2	35.2	*n.a.*	34.6
Municipal elections, 2004/5 (Councils)	80.5	*n.a.*	75.9	69.2

Source: Consejos Nacional Electoral (National Electoral Council – CNE).
* Data belonging to Jiménez and Hidalgo (2014). Penfold (2010) notices that results from of the 2007 Constitutional Reform Referendum have never been officialised by the CNE.

While municipal elections have always been characterised by a low voter turnout (Castellanos Vasquez, 2005; Velasquez Trujillo, 2005), since the end of the Punto Fijo regime and the rise of the Chavez's government, voter turnout rates in national elections have been particularly high[27] – mainly if compared to many other Latin American countries (Figueroa, 2006) – and have "reversed a secular rise [of abstention] over the preceding 25 years" (Helliger, 2005: 19). Table 5.3.2. also confirms that further increasing participation in national elections and referenda is visible in the last ten years. Reduced abstention from voting at the national level, rather than at the local one, is considered a possible consequence of some relevant and successive factors. The first relates to the economic (or distribution) and institutional (or representation) crises that took place during the Punto Fijo regime. Previous governments' ineffectiveness in providing basic services, people's disapproval for the tentative adjustment programmes, corruption among politicians together with the catch-all nature of the nation's two major political parties (AD and COPEI) produced progressive abstention rates in the 1980s and 1990s whilst, at the same time, a renewed interest in presidential elections aimed at sustaining the radical changes proposed by Hugo Chavez (Levine, 2002; McCoy and Meyer, 2004).

A second factor rests on Chavez's efforts and capacity to mobilise the citizenry – especially the poor masses – in order to increase the electoral consensus over his alternative political proposal to the previous regime. The government's economic and social programmes (*misiones*) aimed at providing services to the urban and rural poor and the resulting benefits for the popular sectors in terms of education, health, housing and land redistribution have clearly played a central role (McCoy and Meyer, 2004; Helliger, 2005; Figueroa, 2006; Maingon, 2009; Penfold, 2010). Motivations for citizens to go to the ballot box to sustain Hugo Chavez have been linked, on the one hand, to the people's genuine support of the *misiones*. Figueroa (2006: 207), in particular, underlines the tight positive connection between national electoral participation and economic inclusion of the poor strata which has occurred in Venezuela: "[t]his new dynamic of inclusion into the national *economic* agenda has been a clear component of this new *political* inclusion". On the other hand, the Chavez government's ability to include and equally involve in public policies the many people traditionally excluded from the decision-making and redistribution policies together with the informal labour sector has helped reducing abstention in national elections (Wilpert, 2002; Figueroa, 2006; Helliger, 2005): starting from the end of the last century, there is a enhanced positive correlation between community

27 Voter turnout in the national elections since the first Chavez mandate was: in 1998: 63.5%; in 2000: 56.3%; in 2004. 69.7% and in 2006: 74.7% (Consejo Nacional Electoral, www.cne.gov.ve).

participation and electoral participation (Briceño, 2009). However, some other critical literature argues not only that Chavez has only made people feel as though they were included and involved in public polices (Canache, 2004; Gil Yepes, 2004; Smilde, 2011), but that, eventually, "the *misiones* were directed to low-income sectors not only with a redistributive purpose but also for the votes' direct purchase" (Penfold, 2010: 34), thus feeding clientelist relationships (Maingo, 2009; Reyes, 2013). Such a "cult of personality" is deemed to possibly jeopardise the sustainability and viability of participatory programmes like the missions (Penfold, 2010, Lopez Maya and Lander, 2011; Córdova Jaimes, 2012).

Finally, Maingon (2009) advances the hypothesis of the polarisation of the electoral basis. In this perspective, participation in national elections and referenda increases as abstention from people opposing the government could ultimately reinforce the actual power system, preventing alternative forces to emerge. This hypothesis seems supported by the much higher rise in opposition votes with respect to the actual government, between the presidential elections of 2006 and 2012 (Ponce and Rangel, 2013).

Participative democracy

When looking for the "existence of a mechanism to facilitate participatory planning" (UN-HABITAT, 2004: 38; UGI indicator No. 17) in Venezuela, whether this related to water or not, the answer cannot overlook the unicity of this country with respect to community participation, and the highly polarised debate about representative and participatory democracy (Penfold, 2010; Goldfrank, 2011; Smilde and Helliger, 2011). A comprehensive discussion of participatory democracy in Venezuela and the large debate it generates is neither possible nor appropriate here, but outlining some key elements will help us to contextualise local governance arrangements for improved water services.

Although, from a normative perspective, the first step towards the institutionalisation of participative democracy is represented by the Reform of the Organic Municipal Government Act (*Ley organica de Regimen Municipal*) of 1989 (Lopez Valladar and Añes Hernandez, 2005), the current Venezuelan Constitution (1999) is widely considered as the normative reference document promoting and supporting "participatory democracy" not only for consultation and a project's practical execution, but also for creating a new management and organisation of public affairs. With respect to the previous Constitution (1961), new forms of citizen participation are included in the current charter:

> Participation and involvement of people in the exercise of their sovereignty in political affairs can be manifested by: voting to fill public offices, referendum,

> consultation of public opinion, mandate revocation, legislative, constitutional and constituent initiative, open forums and meetings of citizens whose decisions shall be binding among others; and in social and economic affairs: citizen service organs, self management, co-management, cooperatives in all forms, including those of a financial nature, savings funds, community enterprises, and other forms of association guided by the values of mutual cooperation and solidarity[28] (art. 70).

The juridical re-evaluation of citizen participation in the Venezuelan Constitution rests on these new participatory means, but also on the emphasis given to participation in public affairs, either directly or through their elected representatives (art. 62), and in the political, civic and community life of the country (art. 132); in the selection and denomination of judges, justices and various other public officials (art. 255 and 264); in the achievement of social, family and educational rights (art. 79 and 102). According to Nuñez Nava (2003: 23), the current Constitution differs from the previous one since it gives citizen participation a new role, so relevant as to transform it, within the constitutional mandate, in "an instrument, political request, mechanism, juridical principle and value which build the socio-politic and juridical-philosophical framework of the Venezuelan state".

If we focus on the municipal level, the 1999 Constitution states the "actions of a Municipality within the scope of its competence shall be carried out by incorporating citizen participation into the process of defining and managing public affairs, and monitoring and evaluating the results achieved, in an effective, sufficient and timely manner, in accordance with law"[29] (art. 168). The 2005 Organic Municipal Public Power Act (*Ley Organica del Poder Publico Municipal* – LOPPM), on its side, stresses the citizen's right to participate in public management and the municipal obligation to guarantee citizen involvement in decision making processes (art. 253-279)[30]. Public participation in the development, review and control of municipal policies and service management should be performed through the *Consejos Locales de Planificacion Publica* (CLPP). The CLPP might include representatives of a wide number of community-based organisations and local citizen associations working towards improving living conditions (UGI-indicator 17: "Public Forums"), such as the *asociacion civil*, *asociacion de vecinos*, *comité de tierra urbana (CTU)*, *mesa tecnica de agua (MTA)*, *mesa tecnica de luz* or *comité de energia*, *comité de educacion*, *cooperativas vecinales*,

---

28 Unofficial English translation from the website: www.embavenez-us.org.

29 *Idem*.

30 Article 275 of the 2005 LOPPM was further modified by the 2010 reform of this law to include the new forms of citizen organisation, such as the *Comunas*.

comunas and *consejos comunales*. According to the *Ley de los Consejos Locales de Planificación Pública* (LCLPP, 2002: art. 8), these councils have to:

> promote the network of parishes and communal councils in every space of civil society that, in general, respond to the nature of the municipality, whose function will be to act as the main centre for participation and *protagonismo* of people in the formulation, execution, control and evaluation of public politics, as well as to help organised communities to present ideas and proposals to the CLPP.

Some scholars stress that the role of the CLPP has been progressively challenged for a number of different reasons. First, mayors were not willing to activate them, transfer resources to them or wanted to exercise control over them (Goldfrank, 2011). Second, CLPP were somewhat autonomous and not totally subordinated to the national power (Rangel Guerrero, 2010) so that participation at a micro-level rather than at the municipal level was increasingly promoted through the communal councils (Goldfrank, 2011) while the relative laws (*Ley Organica de los Consejos Comunales*, 2007 and 2009) denied article 8 of the LCLPP (Peraza, 2012). Third, the Communes Law (*Ley de las Comunas*, 2010) marked the rejection of the municipality as the primary political unit of the national organisation, in favour of a parallel project of a communal state, bypassing participation at the municipal level (Brewer-Carías, 2011).

A part of the literature acknowledges that the current scope for a participation *protagonica* (self-reliant) in Venezuela has generated a strong sense of satisfaction with democracy and confidence in the improvements of living conditions amongst citizens, particularly within the poorest (Wilpert, 2005; Gibbs, 2006; Helliger, 2011; Smilde, 2011). Gibbs (2006) shows that public participation in decision making in Venezuela has contributed to making this country an exception in terms of its citizens' satisfaction with how democracy works, amongst the other low scores in Latin America. Based on his analysis of Venezuelan neighbourhood associations and local committees working in the health sector[31], this author argues that their involvement in the service delivery "represents a process of *re-politicisation* from the community to the national level" (Gibbs, 2006: 269). Furthermore, the government's social policies are deemed to have contributed to the recognition of the ignored spaces and hidden histories in informal settlements (Smilde and Hellinger, 2011). In particular, citizen participation is supposed to "have

[31] Gibbs (2006: 273) explains that neighbourhood associations and local committees working in the health sector "are responsible for identifying their community's needs, for contributing to designing programmes to solve local problems and for evaluating programmes".

allowed many Venezuelans, but especially the poor, to feel included in Venezuelan democracy more than they ever have been before", while "many people who live in the barrios consistently report that their sense of hope and of being noticed by the government has increased tremendously" (Wilpert, 2005). One professor at the Central University of Venezuela expressed her opinion on the positive effects of self-reliant participation by saying: "I don't know if this government is good or bad, but now people do dream!"[32].

However, these unquestionably positive effects are accompanied by criticisms of the viability of the proposal. Some major limitations may occur in the practice of participation in the Bolivarian Project. First of all, supporting a self-reliant participation can be in contradiction with the other normative aspects emanating from the 1999 Constitution (Cuñarro *et al.*, 2006; Levine and Romero, 2004; Brewer-Carías, 2010; Córdova Jaimes, 2012). Cuñarro *et al.* (2006: 131) argue for instance that the democratisation of the public administration through citizen involvement in the processes of decision, implementation and control of public affairs clashes with the increasingly strong and central state defending citizens' rights and social purposes: "How can responsibilities be delegated and resources decentralised while maintaining the State centrality, as requested by its 'social' character?" In the same vein, Córdova Jaimes (2012: 680) wonders: "How can be matched mechanisms aimed at the protagonic participation of citizens … with a centralised model of public planning and management?" This paradox unfolds at two levels. Citizen participation might be restricted to public decision-making, but excluded from controlling public administration (see also Arconada, 2005); or, the public administration could encounter a "democratic restriction"[33] through the diminished capacity to produce goods and services as a result of the burdensome administrative procedures related to participative democracy and by citizen requests (Cunill, 2004; Cuñarro *et al.*, 2006; Córdova Jaimes, 2012).

Second, the practice of participative democracy in Venezuela has been deemed to have possibly faced internal obstacles given the potential community organisations' dependency on the strong personality and image carried by the former Venezuelan President Hugo Chavez. Wilpert (2005), for instance, underlines that the more important negative outcomes of Chavez's "cult of personality" are represented, on the one hand, by the potential loss of citizens' independent critical thought because of their complete acceptance of the leader's discourse and, on the other hand, by

32 Personal interview, 2006.

33 Cuñarro *et al.* (2006: 127) define a democratic restriction as "the limitation of political decisions caused by the requests from most of the electorate".

the necessity to be part of the "president followers" (*chavistas*) in order to benefit from the many government social programmes. With respect to the latter aspect, Brewer-Carías (2010: 380) accurately demonstrates how "only the government parties and the supporters of the president are 'represented'; the opposition parties and organisations are completely excluded or marginalised from political life". Other critics wonder if civil society can organise itself effectively and on a long-term basis outside the current politics without falling to dependency on their leader, as it was before the Chavez Republic (Levine and Romero, 2002, 2004). This concern seems particularly appropriate if we look to the most recent reforms concerning "popular power" or "citizens' power", with the introduction of the communal councils (2006 and 2009) and the communes (2010), and the way in which these citizen assemblies are directly dependent on the executive power (Brewer-Carías, 2010, 2011; Penfold, 2010).

Third, participatory democracy may encounter serious restrictions if decentralisation is not accomplished. As Brewer-Carías (2010: 280) cleverly indicates

> For democracy to be participatory, in addition to being essentially representative, it must allow citizens the possibility of participating in public affairs, which is possible only when they have access to power. This is possible only when power is near to citizens, which necessarily implies the presence of a well-established, well developed autonomous local government in every locality and urban or rural settlement. This means that political participation can be founded only on political decentralisation, through the creation of autonomous political entities that permit local self-government.

According to Brewer-Carías's analysis of democracy in Venezuela, the absence of local territorial autonomy and decentralisation – mainly because of power concentration, as previously discussed in this chapter – may turn the potential for genuine democratic participation into simple popular mobilisation.

If we now look to the number of "civic associations" officially registered in local governments (as per UGI indicator 18), as previously indicated, no official data is available for the two selected municipalities of Paz Castillo and Cristóbal Rojas: as in most Venezuelan municipalities, it is not common for community-based associations to register whilst, at the same time, it is extremely difficult to find civic associations' registering. The reason lies in the gap between the legal framework for citizen participation and the way citizen participation really operates on the ground. Actually, community associations, committees and organisations are generally formally regulated by Venezuelan legislation with respect to their establishment and the way they relate to other formal institutions, in order to "give institutional form to popular participation and raise it to a higher level" (Figueroa, 2006: 209)[34]. However, it often occurs that community-based associations begin to organise themselves and operate outside of the formal regulations applying to them – developing informal practices and informal arrangements and relationships with other stakeholders – whilst at the same time encountering institutional limitations in accessing decision making. It is interesting to note that the Local Public Planning Councils law (LCLPP, 2002) has been extensively contested by the Venezuelan community sector as, to enable the community to participate in the programmes and actions taken by the municipality, it was necessary to be an "organised community", which means being formally constituted and legally inscribed on the official register. As a consequence, the 2006 reform of the LCLPP law has changed this requirement, making conditions for communities to participate in the LCLPP less strict (Asamblea Nacional de la Republica Bolivariana de Venezuela, 2006).

With respect to the water sector, INDIA (2006: 2) underlines how community participation in the delivery and organisation of service provisions represents one of the two fundamental characteristics of the current changes in improving access to drinking water, together with the decentralisation process. Within the norms brought in by the LOPSAPS, there is a wide consensus on the importance of participatory programmes – represented by the *mesas tecnicas de agua* and *consejos comunitarios de agua* – as tools for promoting and supporting equitable and effective access to drinking water. Amongst the main changes and contributions brought about by MTAs in building more sustainable access to and organisation of water services, critics list the MTAs' capacity to improve access to water for low-income dwellers, to develop a new culture promoting environmental conservation and water payment, to participate in and negotiate for the delivery of water services with public institutions (Cariola and Lacabana, 2005b; Hidroven, 2005; Arconada, 2005; Francisco, 2005; INDIA, 2006; Lopez Maya, 2008; Moretto, 2010; Lopez Maya and Lander, 2011; McMillan *et al.*, 2014).

There are no recent detailed figures for MTAs operating within Hidrocapital water jurisdiction. Higuerey *et al.* (2012) showed that MTAs located in the CMR, Miranda and Vergas states were 553 in 2006, representing almost 20% of all Venezuela MTAs, but "only" 712 in 2010, accounting for less than 9% of the total number (see table 2.2. which

---

34 For example, the *Ley Orgánica para la Prestación de los Servicios de Agua Potable y Saneamiento* regulates the responsibilities for and relationships with the *mesa tecnica de agua* and other formal institutions (Asamblea Nacional de la República Bolivariana de Venezuela, 2001), as well as the *Ley de los Consejos Locales de Planificación Pública* (Asemblea Nacional de la República Bolivariana de Venezuela, 2002) and the following reforms in 2006, 2010 and 2013.

provides an estimation of the MTAs' evolution and distribution). While at the national level, the Ministry of Citizen Power for Environment estimates that 9,700 MTAs have been constituted in Venezuela up to 2013[35] (impressively quadruplicating in seven years), at the municipal level, figures are not available. So, with respect to the two selected municipalities, it is not possible to get an estimate of the number of MTAs.

Vibrant and widespread citizen participation in water provisions, however, is not always accompanied by the similarly equitable and active involvement of all the stakeholders who should be included in the service delivery process, according to the LOPSAPS. Our analysis reveals that, in effect, in the two selected municipalities, the practice and local dynamics of community participation for improving water services is largely characterised by the municipality's deficient knowledge concerning the existing MTAs and by a structural lack of communication between these two actors, thus confirming that the municipality is severely excluded from these participatory programmes. Also, water service improvements are conceived and implemented by frequent and more informal relationships and information exchanges between the community and the state-owned water company, without any fundamental role being played by the local government (see next chapter for further details). Within these common characteristics, some minor differences apply nonetheless between the two municipalities.

In Cristóbal Rojas and Paz Castillo, all the interviewed municipal public servants affirmed that they did not have any particular contact with the technical water committees in their municipalities, while in Cristóbal Rojas they also confirmed that they did not even know the status of the MTAs in their municipality[36]. They received much more information about water related problems through their meetings with Hidrocapital representatives rather than directly through the community. The director of "Urban and Rural Planning" in Cristóbal Rojas stated: "I am engaged with the community – and the community knows me – but I have never been called by an MTA ... if I want to know something about the MTA, I ask Hidrocapital, this is how I get informed, but it is not the community that tells me". No regular and formal meetings between the technical water committees and the municipality are organised in order

35 Information obtained from the Ministry of Citizen Power for Environment website (http://www.minamb.gob.ve, accessed on the 25th of March 2014).

36 A single exception was found in the municipality of Paz Castillo, where the mayor was deeply involved in the provision of drinking water delivery. Nevertheless, his involvement was not assessed in respect to all the MTAs created in the municipality but just against the community selected for this research (*Hacienda el Carmen*), where he showed a great knowledge of the MTA history and development and also a deep practical and personal involvement in helping the community to get water.

to discuss water related problems. In general, municipal officers said that when community members have gone to municipal offices to complain about water shortages, they suggested contacting Hidrocapital directly in order to establish an MTA or to organise the technical water committee's activities with the hydrological company. All interviewed municipal public servants acknowledged that communication between the municipality and the water company concerned just two main subjects: information about communities struggling with water delivery problems and the operational and technical solutions to be arranged between the municipality and Hidrocapital with respect to the piped network's construction.

Despite its central role in the delivery of drinking water provisions from a normative perspective, the municipality appears to be excluded from the local governance system for improving service delivery and from any form of involvement in MTA constitution or development. Information, communication, agreements and collaboration for improving water provisions are mainly organised and put into effect by the community and Hidrocapital. The municipality's role is thus merely in providing institutional support to the arrangements made between the community and the water company. For example, the Mayor of Paz Castillo affirmed that "the municipality has above all a role of facilitator … we are useful as facilitators and at the economic level when the community cannot afford the costs", while the ex-community coordinator of the Losada Ocumarito system stressed that "the Mayor is simply a presence, a person who represents people but without deciding … The Mayor and the parish councils (*junta parroquial*) are just managers (*gestores*) of public policies, not deciders …". If municipal public servants are generally not aware about the technical water committees' activities and development, their role in supporting and facilitating community-Hidrocapital collaboration clearly looks extremely doubtful.

Finally, the previously outlined potential limitations and obstacles to participative democracy seem to also be reflected in the water sector. First, incomplete decentralisation – in drinking water provisions too – reduces possibilities for the municipality to play a fundamental role in the local governance arrangements to improve water services. Second, as a consequence, citizens participate in the decision-making process for, and operationalization of, service delivery improvements, not at the municipal level but via the central power institutions represented by the state-owned water company. Third, the absence of institutionalised, formal and regular meetings amongst the main stakeholders who should be involved in the water delivery process (notably the municipality, the MTA and the hydrological company) help communities to directly and swiftly access the state-owned water company. These three elements reinforce MTAs' direct relationships with state institutions and, together with MTA participation at the regional and national water meetings, they increase a deep sense of belonging to, but also dependency on, the state and the central power.

To conclude, the MTAs participation in service delivery has unquestionably contributed to improving drinking water supplies for low-income communities and to the mobilisation of these communities in order to develop better access to basic services and the decision-making process. Nevertheless, in the case of complete decentralisation of water services to municipalities – as per the current LOPSAPS – certain risks could still exist. The practice of a participative democracy based on strong linkages between MTAs and state-owned water companies together with weak or non-existent relationships between the community and the municipality, as assessed in the two selected municipalities, could easily jeopardise the viability of this proposal.

5.4. Accountability defied by weak municipal autonomy

Within the wide matter of accountability in local governance, we focus here on the criterion represented by downward accountability, which refers to the "accountability of the local government to civil society" (UN-HABITAT, 2004a: 23). Upward accountability – from local institutions towards higher levels of government – is not considered here, given that the emphasis of our analysis is concentrated on the very local aspects of governance arrangements, at the municipal and community levels. In this perspective, UN-HABITAT identifies four main features characterising the accountability principle and define the 7 accountability UGI indicators with respect to these features: transparency, responsiveness, integrity, and disincentives and protection against corruption (see table 5.4.1.).

The accountability sub-indexes of Paz Castillo and Cristóbal Rojas present a wide discrepancy: this is due to the absence, in Paz Castillo, of two specific municipal bodies requested by the current Venezuelan Citizen Power legislation, but already established in Cristóbal Rojas at the time of field investigation: the Municipality Comptrollership Office (*Contralória Municipal*) and the Citizen Office (*Oficina de Atencion al Ciudadano*). As a consequence, the municipality of Paz Castillo did not present a formal body to provide citizens with an official report regarding contracts, tenders, budgets and accounts (UGI indicator No. 19), it was not equipped with facilities established within the local authority to respond to complaints and to receive information on corruption (UGI indicator No. 22), to investigate and report cases of corruption (UGI indicator No. 23) and to run a regular independent audit of municipal accounts (UGI indicator No. 25), (see also Annex 3 on accountability indicators calculation for further details).

Table 5.4. UGI accountability indicators and main features defining accountability in local governance

ACCOUNTABILITY UGI INDICATORS				***ACCOUNTABILITY MAIN FEATURES***
No.	**Indicator**	**Data (X)**		
		Paz Castillo	**Cristóbal Rojas**	
19	Formal Publication (FP)			Transparency
	a. Formal publication: contracts and tenders (CT)	*No*	*n.a.*	
	b. Formal publication: budget and accounts (BA)	*No*	*n.a.*	
20A	Control by higher Govt. (CG)			Responsiveness
	a. Control by higher Govt.: close local government (CLG)	*No*	*No*	
	b. Control by higher Govt.: removal of councillors (RC)	*No*	*No*	
20B	Local government authorities (LGA)			Responsiveness
	c. Local government: set local tax levels (SLT)	*Yes*	*Yes*	
	d. Local government: set user charges for services (SUC)	*No*	*No*	
	e. Local government: borrow funds (BF)	*n.a.*	*n.a.*	
	f. Local government: choose contractors for projects (CP)	*Yes*	*Yes*	
21	Codes of conduct (CoC)	Yes	Yes	Integrity
22	Facilities to receive complaints for water (FRC)			Integrity
	a. Official appointed to receive complaints on public authorities (OA)	*No*	*Yes*	
	b. Exclusive facility to receive complaints on corruption (EF)	*No*	*Yes*	
23	Anti-corruption commission (ACC)	No	Yes	Corruption
24	Personal Income and assets (PIA)			Corruption
	a. Disclosure of personal income and assets (PIA)	*Yes*	*Yes*	
	b. Disclosure of family's income and assets (FIA)	*n.a.*	*n.a.*	
	c. Income and assets regularly monitored (IAM)	*n.a.*	*n.a.*	
25	Regular independent audit (RIA)	No	Yes	Corruption
	Accountability sub-index	**0.28**	**0.85**	

For those indicators concerning the provision of public services (UGI indicators No. 20B and No. 22), the investigation has also taken into account Hidrocapital's accountability to citizens. Anyway, the calculation of these UGI indicators only focuses on municipal performance, thus confirming that the provision of water services had not yet been decentralised as the water company was still responsible for the service provision, user charges' definition and complaints' collection. Especially in the case of Cristóbal Rojas – which scores very high – the overall result for the accountability sub-index seems to only partially reveal the local governments' limited accountability to citizens with respect to the provision of water services.

Downward accountability: accountability of municipalities to civil society

One of the main legacies left by the Punto Fijo regime is the widespread desire to make the state and public institutions less paternalistic and more accountable to ordinary citizens, in order to regain people's confidence in and support for the political system (McCoy and Myers, 2004; Gil Yepes, 2004; Smilde, 2011). In the meantime, for the ideological policy orientations belonging to the Bolivarian Project (redistribution, anti-neoliberalism, participatory democracy and inclusion) to be pursued, it very soon became clear that the Chavez government should have strengthened accountability and thus fought corruption and inefficiency (Wilpert, 2002). A great number of laws and norms have been thus produced since the 1999 Constitution, addressing the issues of transparency, participation and control in public management (Delgado, 2004; Brewer-Carías, 2010).

Venezuelan legislation provides specific regulation for all of the UGI indicators belonging to the accountability principle. The Anti-corruption Law (2003, *Ley contra la Corrupcion* – LCC) and the Organic Law on the Republic General Comptrollership and National Fiscal Control System (2001, *Ley Organica de la Contraloria General de la Republica y del Sistema Nacional de Control Fiscal* – LOCGRSNCF, further modified in 2010) represent the main normative sources that address accountability indicators. Formal publication of information regarding contracts, tenders, budgets and accounts through a detailed report (UGI indicator No. 19) is compulsory for all public institutions according to the LCC (2003: art. 19), as well as the disclosure of personal income assets for locally elected officials (UGI indicator No. 24; LCC, 2003: art. 23), a local agency to investigate and report cases of corruption (UGI indicator No. 23; LCC, 2003: art. 41) and a regular independent audit of municipal accounts (UGI indicator No. 25; LOCGRSNCF, 2001: art. 42 and 57). Codes of conduct aimed at ensuring local government commitment towards the integrity of its officials (UGI indicator No. 21) are stated by the LCC (2003: art. 6) but also incorporated into other government regulations at the state level, such as the *Ley del estatuto de la function publica* (2002), the LOCGRSNCF (2001, 2010) and the *Ley Organica del*

Poder Pùblico Municipal (2005). The existence of a facility established within the local authority to respond to complaints and a local facility to receive complaints and information on corruption (UGI indicator No. 22) is ensured by the Citizen Office (*Oficina de Atencion al Ciudadano*).

Despite this national comprehensive normative framework, there is a considerable difference between the accountability sub-indexes of the two municipalities because of different scoring in the calculation of some specific UGI indicators (No. 19, 22, 23, 25). These differences rest principally on the existence or absence of two local agencies belonging to the Republic General Controllership: the Municipality Comptrollership Office (*Contralória Municipal*) and the Citizen Office (*Oficina de Atencion al Ciudadano*):

- The Municipality Comptrollership Office is in charge of the "control, observation and inspection of incomes, expenditures and municipal goods, as well as the operations related to these ones" (CRBV, 1999: art. 176; LOPMM, 2005: art. 100); the *Contralória Municipal* was already regulated by the Organic Municipal Government Act (1989, *Ley Organica de Regimen Municipal*) but with the 1999 Constitution, the Controllership has become one of the bodies belonging to the Citizen Power, together with the ombudsman facility (*Defensoria del Pueblo*) and the Public Ministry (*Ministerio Publico*) (Rojas Hernandez, 2002).
- The Citizen Office is a compulsory agency which has to be established within all of the state bodies and institutions – and thus the municipalities too – in order to inform citizens about the use of public goods and public resource expenditure, provide a detailed report on the public patrimony's administration available to all people, and collect and process citizens' complaints and suggestions (LCC, 2003: art. 9 and 10). Transparencia Venezuela (2004: 7) highlights how the *Oficina de Atencion al Ciudadano* is "a room for accountability and social controllership … aiming at realising the right to access to Public Information".

The Cristóbal Rojas municipality is equipped with both these local agencies within the same structure, the *Oficina de Participación Ciudadana y Contraloría Social de la Contraloría Municipal del Municipio Cristóbal Rojas*. Conversely, at the time of investigation, the municipality of Paz Castillo had set up neither a Municipality Comptrollership Office nor a Citizen Office. As a result, the former municipality scores much higher than the latter.

Drawing on the considerations made for the decentralisation process, it appears that the current complete and precise legal framework addressing the issues of transparency, fiscal control, integrity and corruption, does not necessarily correspond to the real execution of these functions as the municipal capacity for citizen accountability depends, above all, on its capacity to set up the local bodies charged with the aforementioned responsibilities. The reasons motivating the absence of the *Contralória Municipal* in Paz Castillo are diverse. According to its mayor, "the main obstacle in the setting up of the Municipality Controllership Office has been the lack of resources, as the priority has been given to communities' necessities rather than to the bureaucratic expenditures that this body would generate; it is nevertheless clear that this office shall confer much transparency to the municipal administration" (*La Voz*, 2005). While according to the municipal financial director, the setting up of the Municipality Controllership Office is a long process that takes considerable time to be realised.

With regards to the UGI indicators related to basic service provision – and in particular to the provision of water services (indicators No. 20 and No. 22) – the normative framework aimed at ensuring municipal accountability in relation to the delivery of public services cannot be fully applied to the water sector, as long as water services in the Miranda State are still provided by the state-owned water company. We point out here two examples. First, in both the Paz Castillo and Cristóbal Rojas municipalities, as in all Venezuelan local governments, municipalities could in theory set local tax levels, choose contractors for projects – according to the procedures established by national legislation – and set user charges for services[37] (UGI indicator No. 20B), thus holding quite large political and legal autonomy and authority in organising the delivery of basic public services. Nevertheless, the possibility of the local government to set user charges for services, depends in reality on whether these services are provided and managed by the municipality or an external agency (public or private), as in the case of drinking water provisions. User charges for water are, in fact, currently set and collected autonomously by Hidrocapital, without any say from the local government. In the same way, the LOPSAPS (2001, art. 83) establishes that water suppliers have to calculate tariffs on the basis of the "tariff model" set up by the National Superintendency of Drinking Water and Sanitation Services, where the Superintendency reserves the right to object if the tariff model is not correctly applied. The municipal authorities only have the option to reprove the suppliers' tariffs in the case of inadequacy within the tariff model (LOPSAPS, 2001, art. 83, *paragrafo unico*): this

[37] Interviews with public officials in the technical services of Paz Castillo and Cristóbal Rojas did not provide an answer on the possibility for municipalities to *directly borrow funds* for their local projects without permission from higher governments. It seems that the two municipalities have never faced the problem of borrowing funds from external agencies.

condition, again, strongly reduces the autonomy of local governments in managing water services in favour of central state institutions and, at the same time, it also reduces municipal accountability to citizens with respect to the delivery of basic services.

Second, a similar consideration can be outlined with respect to the indicator concerning the existence of facilities for citizens to complain about their water provisions' flaws and shortcomings (UGI indicator No. 22). According to our interviews with municipal public servants and Hidrocapital officials, even when the municipality has established the Municipality Comptrollership Office and/or the Citizen Office (as in the case of Cristóbal Rojas), citizens' complaints are not directly processed by the municipality but just passed on to Hidrocapital services. Fortunately, along with our fieldwork results, citizens are aware that they may contact the water company directly as it has organised many commercial offices (*oficinas comerciales*) to collect drinking water payments and complaints about the service's management and provision. In addition, Hidrocapital receives complaints through their easy to remember hotline 0800-7682253 (0800-potable)[38].

To conclude, municipal accountability towards citizens in Venezuela seems to benefit from a two-fold enforcement. On the one hand, the compulsory setting up of specific municipal bodies (the *Contralória Municipal* and the *Oficina de Atencion al Ciudadano*) aims at formally ensuring municipal responsiveness, transparency and integrity towards citizens while, on the other hand, the complex structure of citizen participation through the various institutional forms (mainly represented by the Public Planning Local Councils, CLPP – *Consejos Locales de Planificacion Publica* but also by the current CBOs such as the MTA) tends to establish a bottom-up mechanism able to foster municipal downward accountability.

Nevertheless, both these mechanisms for improved accountability appear at least partially to be failing on the ground. In the case of municipal bodies for transparency, responsiveness and integrity, we have just showed how the incomplete (water service) decentralisation to municipalities makes these bodies – provided they exist – ineffective with respect to drinking water management. This situation clearly raises the issue of who should be accountable to citizens for the delivery of equitable and efficient water services given the inconsistency between the responsibilities stated at both the legal and operational levels.

38 Nonetheless, according to interviews with Hidrocapital officials, it seems that citizens' complaints can be addressed to Hidrocapital only if the community or the person making the complaint is already paying for the water service.

In the case of an enforced municipal accountability through citizen participation, the effectiveness and sustainability of these mechanisms cannot be taken for granted. If "the broader significance of participatory democracy is that it addresses social exclusions and incomplete citizenship through the increased *accountability* of elected officials to the citizens by directly involving the latter in the political arena, redistributing power by expanding popular deliberation and participation in policy areas and redirecting revenue to address the concerns of the people" (emphasis added; Wagner, 2004a), this seems to not involve the municipal sphere. In reality, although "local citizens' assemblies enjoy a constitutional status, so that they can force local officials to be more accountable" (Wilpert, 2005), municipal servants are systematically bypassed. As we previously indicated, the CLPP – which should constitute the formal citizen assembly for direct participation in municipal affairs – have been progressively disregarded in favour of other forms of community organisations, such as the community councils or the communes, which are much more autonomous from municipalities and linked to the central rather than local powers. The next chapter will explore the functioning of two MTAs and explain the architecture of their relationships with state institutions and local representatives.

CHAPTER 6

Low-income Communities' Involvement in Local Governance for Improved Water Provisions

Let's now move to the local governance arrangements within low-income communities, to see how governance for improved urban water provisions works in practice. The principle framework of the Urban Governance Index, and the relative indicators, have been translated in a qualitative methodological approach at the settlement level, as presented in chapter 4. Unlike the assessment of urban governance at the municipal level, through quantitative indicators and factual data, the boundaries between the four UGI principles are even less clear-cut and easily circumscribable at the community level. This makes it quite challenging to elaborate a precise and unequivocal description of how these principles might work at the settlement level, while also featuring cross-references but also possible repetitions. Where applicable, we will avoid useless repeats, rather indicating where each specific issue is analysed in detail in the chapter.

6.1. Effectiveness of urban governance on the ground: case by case

Effectiveness of local governance arrangements amongst the municipality, the water company and the MTA in the delivery of drinking water is assessed against the capacity of formal and informal governance mechanisms developed around the technical water committees to bring about improvements in the effective delivery of safe drinking water, in the enhanced community development within the settlement, and in the responsiveness to community concerns. Results on the ground greatly vary in the two analysed settlements, according to previous water access conditions in each area, the land tenure, the history and structure of the MTA, its capacity to organise itself, the extent of state water company intervention, the commitment of municipal officials, etc. The pages that follow provide a detailed account of these differences, while demonstrating that effectiveness of local governance arrangements is strongly site-specific. Nonetheless, some key trends can be sketched out.

First, concerning the improved delivery of drinking water: if in the *Paso Real 2000* settlement (PR2), residents experience different levels of water provision inside the area and consider their access conditions inefficient, in the *Hacienda el Carmen* (HeC), local institutions (the municipality and Hidrocapital) and the majority of community residents recognise that the service upgrade realised through piped connections in the whole settlement is efficient and works well. Wider improvements to the water network in the latter settlement are certainly linked, on the one hand, to their worse previous water access conditions, which caused greater community mobilisation and consequently a timely response by the institutions to solve water shortages and, on the other, to the less constraining land tenure situation, as the area belongs to the public domain (see also chapter 4). They seem also associated with a limited awareness from community residents concerning the institutional mechanisms developed around the technical water committee.

Second, with respect to the MTAs influence on enhanced community development, if it is clear that the technical water committees are not an isolated experiment to improve service provisions but a part of a wider national project for a more participatory democracy, greater and effective collaborations between the MTAs and other CBOs cannot be taken for granted. Our two examples show that, in one case (HeC), the positive outcomes of the technical water committee in terms of water access improvements have been important in promoting new community-based associations in the settlement in order to gain access to other services and facilities. Conversely, in the other case, there does not seem to be much cooperation between the MTA and other community-based organisations, mainly because of the very poor cohesion within the technical water committee, and the scattered results it obtained.

Third, effectiveness in responding to community concerns on water provisions seems to be connected to both an effective water delivery system and an effective communication mechanism between the MTA and local institutions. The ineffectiveness of local governance arrangements in the PR2 settlement in providing a good water service for the whole community, for example, is without a doubt due not only to the lack of a comprehensive project by Hidrocapital, but also to the limited and imprecise knowledge that local institutions have about the water access conditions of this settlement.

Finally, a direct relationship between effective improvements in water delivery and the establishment of a formal and well recognised technical water committee does not seem to exist. In the HeC, the water project has been successfully realised without any particular awareness from community members of the link between MTA activities and the water-networked project. Conversely, in the PR2 settlement, the efforts of the technical water committee leaders in building up an official and well organised MTA (see chapter 4) have resulted in just partial water company intervention in one terrace (which is nevertheless identified by residents as a result obtained through the MTA commitment). Most water provision improvements in the latter settlement rest on the informal practices among community members, without the support of the MTA, Hidrocapital or the municipality.

Effectiveness in the delivery of safe drinking water and in enhancing community development

Improved water provisions and community awareness

When looking to governance arrangements for improved service provisions on the ground, the existence of a concrete upgrade in service delivery is essential to appreciate the effectiveness of these arrangements, but not sufficient. Considering whether there exists awareness among residents concerning the institutional and community based mechanisms and processes that led to these improvements appears to be a fundamental aspect to understanding how governance works in practice. Perhaps surprisingly, when water delivery is effectively improved for everybody, most of the community is not aware of the underlying process.

In the *Hacienda el Carmen*, all stakeholders involved in the local governance process acknowledge the improvements in water accessibility. Interviewed community members (Q1)[1] and MTA leaders affirm that the new network represents a significant improvement in accessing water, with respect to the previous illegal connections or water tanker trucks. Despite few people highlighting their aspiration for a 24 hour service, the residents show their satisfaction with the current system and the timeframe of water distribution – one day out of two (Q11). Anyway, if it is doubtless that community members consider themselves as being much better off with the new pipeline network, they nevertheless link this improvement in water access to the collaboration between the community and Hidrocapital, without any particular reference to the technical water committee or condominiums (Q3). The link between the project's realisation and the MTA or condominium role in obtaining it, it is in effect rather indefinite. In one of the settlement's sectors (*Araguaney Amarillo*), very few people knew that the pipeline was installed with the support and commitment of the MTA and water condominiums, while those remaining considered the network to be the result of a Hidrocapital operation carried out with the community's labour contribution. Likewise, in another sector (*Colinas*), the majority of the interviewees associated

1 All responses to the 11 questions used for the fieldwork with community dwellers are presented in Annex 2. Questions are here referred in the main text through the letter Q and the number of the relative question in Annex 2.

the presence of water condominiums with payment collection, rather than with the implementation of the whole water project.

For MTA leaders however, the role of the technical water committee in the construction of the water network is clear and fundamental. In their view, this result was achieved through joint work between the MTA, the community, Hidrocapital and the municipality, but also through the strong leadership of the water technical committee members. Besides providing food and accommodation to Hidrocapital engineers during the works, they affirmed that they were considered "consultants" and "technicians" at the same level as the water company's engineers. One leader of the *Araguaney Amarillo* sector said that:

> The engineer who worked in this area considered me as a colleague of the project; he said that I was the person responsible for calling Hidrocapital. We worked together, everyone putting in the network construction our specific knowledge ... We never went to Hidrocapital to create problems but to find solutions, and this is the idea of the MTA: to show a problem but, at the same time, to find a solution ... and the MTA is regular in doing this work, while the engineers of Hidrocapital are always available; they immediately come if there is a problem, regardless of whether they are paying or not for the service.

From this it clearly emerges how shared responsibilities between the MTA in the HeC and Hidrocapital – in the design and realisation of the water network – have benefited from the informal (and widely illegal) practices and knowledge developed by community residents in getting water in the past, through often unofficial or private arrangements. Indeed, the community's knowledge of the characteristics of the settlement (such as the settlement topography or the location of existing pipelines – legal and illegal) together with those regarding the various informal ways of accessing water (such as the cost of hosepipes materials, ways to store water, the pressure problems related to illegal connections, etc.) provided the basic information upon which Hidrocapital and the MTA were able to create a reliable water project.

A significantly different situation applies to the case of *Paso Real 2000.* As introduced in chapter 4, this settlement has never benefited from a Hidrocapital and MTA water distribution project that included the entire settlement. Although all interviewed MTA leaders recognised that the technical water committee had been successful in obtaining partial assistance from the water company and the municipality, they were still waiting for Hidrocapital to realise a piped network throughout the whole settlement. In the meantime, they had organised their access to water in different ways, ranging from autonomous community resident practices to selective collaborations with the municipality, MTA members, or private consultants. In all cases these practices and relationships were totally informal. For example, in *terrazas* 2 and 7, illegal connections were built in 2002 from a mains water pipe running on the west side of the settlement. This informal network was made with the economic and in kind contributions of community members and the private technical assistance of an engineer (working for Hidrocapital as well); according to the MTA leader of these terraces, residents had good access to water, night and day, all week, except on Thursdays. *Terraza* 8 was provided with an illegal connection to a mains pipe, built entirely by community members without any support from the municipality, Hidrocapital or the MTA; at the time of investigation, half the residents (about 40 households) were lacking water. In *terraza* 10, the piped system was built in 2003 to provide about 45 households with water; according to PR2 MTA leaders, the works were realised through the economic and in kind contributions of community members[2], the support of the MTA in organising community work and in lobbying the local government, the economic contribution of the municipality and – also in this case – the private technical assistance of an engineer, who was also working for Hidrocapital. In the view of the MTA leader of this terrace, not all of the residents had good access to water[3].

The only exception is represented by *terraza* 11 where there has been direct and recognised intervention by Hidrocapital through pressure from the MTA leaders in the area. The piped network was built with the economic support of the municipality, the economic and technical involvement of the water company, and the economic and in kind contribution of community residents. But, community dwellers were not all equally satisfied with the network improvements (Q1). Half of the interviewed residents recognised the significant improvements that Hidrocapital's intervention brought to their area, as they could then access water every day from 9 am to 4 pm, except on Thursdays, through direct household connections or, in some cases, through a small hosepipe fixed to the household connection of a neighbour benefiting from the water company's installation. The other half considered that, since the PR2 settlement was established, there had been no improvements as water had always come in small quantities and with very low pressure, sometimes worsening with the passing of time and obliging residents to rely on a double connection. This latter situation was essentially experienced by

2 According to the MTA leader of terrace 10, the residents' economic contribution was used to buy the hosepipes and to pay for the technical consultancy of the engineer who helped in the works.

3 Information was collected through a focus group with MTA members. Not all the terraces in *PR2* were represented by an MTA member and as a result, specific information is only available for the terraces described here.

people living far from Hidrocapital intervention, essentially located in a limited area of the settlement (see map 10).

Despite these variances in the frequency and reliability of water provisions, almost all the residents (both those benefiting from Hidrocapital's intervention and most of those accessing water through illegal connections) seemed to be satisfied with their water provisions, in respect to their previous situation (Q11). Along the same lines, interviewed Hidrocapital representatives of the Losada Ocumarito system estimated that this settlement was provided with good water service, especially after Hidrocapital's intervention in 2005[4]; while the director of the "public infrastructure" sector said that "as I always go to PR2, I know that they have a good water service but that they have problems with sanitation". The view of the Hidrocapital and municipal officials, who did not consider the water access conditions in the settlement of PR2 critical, together with a certain satisfaction from the residents with available water services, possibly explain why collaboration between the MTA and the water company was at a stalemate.

Finally, whether the role of the technical water committee is acknowledged in the water provision improvements decided upon and implemented by the water company and the MTA, and also in wider settlement upgrades, is also very unclear in this settlement. No more than half of the community members interviewed in *Paso Real 2000*, benefiting – directly or indirectly – from the Hidrocapital project, knew that the improvements in water access had been realised through the organisation of the MTA (Q3), while municipal officials who were interviewed in Cristóbal Rojas[5] affirmed having no knowledge of the MTA and its commitment to improving water access conditions in PR2.

Enhanced community development

In addition to the positive practical results of the MTA works, some water committee leaders in the *Hacienda el Carmen* settlement point out that the committee's task went undoubtedly beyond the construction of the piped system to communicating with the community to ensure more effective collaboration and sustainable use of water services. One of the leaders in the *Araguaney Amarillo* sector highlighted that "the MTA should bring people to respect water as an element essential to a person's life, without wasting it for animals or plants. The MTA should organise

4 Interviews with the community coordinator of the Losada Ocumarito System and the general manager of the Losada Ocumarito system.

5 Interviews with the director of "Urban and Rural Planning", and the director of "Public Infrastructure".

workshops and meetings to teach people not to waste water". Another leader (*El Esfuerzo* sector) underlined that:

> The purpose of the MTA is to obtain the pipeline connection, because previously the police removed all of the illegal connections we had built... In the case where a community achieves all the services, the MTA's work does not stop there: we have to maintain the connection, in order to prevent shortages ... the MTA has also the task to inform the community in relation to the follow up and the timeframe of works to avoid useless complaints... we also have to continue to work because now the problem of the water payment exists ... I think that we should never stop ... it is like a wedding: may death divide them! (*que la muerte los separen!*) (Leader of the *El Esfuerzo* sector).

However, the technical water committee's commitment does not stop at the water sector. For MTA leaders, the achievements of the technical water committee are not only represented by the realisation of the piped network, but also by their capacity to generate other community organisations to lobby for other basic services (such as the *comité de tierra urbana* to deal with urban planning and housing; the *comité de salud* to provide the settlement with health facilities, the *mesa tecnica de energia* to install a power network, etc.). According to one MTA leader from *El Esfuerzo*, "the MTA has to fight to obtain power connections, the sewerage system and street asphalting ... and it can make pressures in order for the works to develop". In effect, this technical water committee held a decisive role in other upgrade operations in the settlement: relying on the experience of the water committee, all of the MTAs of the *Hacienda al Carmen* were organising themselves to carry out the joint work between the community and Hidrocapital to realise the new sewerage system, funded by the Miranda State government. Additionally, in the *Colinas* sector, community members were constructing a health-service clinic with the support of MTA leaders in obtaining funds and managing construction. In general, many water committee leaders were involved in other community-based organisations that were working towards gaining access to other services too.

Curiously, although residents and municipal public servants were not particularly aware of the water committee's purpose[6], the important role that the MTA carries out with regard to both the physical realisation of

6 Municipal engineers who were interviewed in Paz Castillo were not aware of the technical water committee's results. They explained they were particularly involved in water distribution in rural settlements by means of two municipal tanker trucks that distributed drinking water for free, thus being aware that some private truck tankers were still operating in the area of the HeC. Poor knowledge of the water committee seems thus much more linked to the lack of specific information about recent progress in the delivery of drinking water in this settlement.

the piped network and the further development of other community-based organisations was well acknowledged by the mayor[7] and Hidrocapital officials. They confirmed in fact that the MTA in this settlement represented a great example and the starting point to work out the concept of *participacion protagonica*, promoted by the government since 1999.

Also in *Paso Real 2000*, despite MTA leaders affirming that the technical water committee had not yet made Hidrocapital realise a piped water network throughout the whole settlement, most of them argued that the water committee could and should keep on lobbying the institutions, with the support of the other PR2 community-based associations: "The MTA's future contribution could rest on lobbying the company – together with the *Comité de Tierra Urbana* – in order to keep the project, to continue with the MTA and the *Comité de Tierra Urbana* together ... to do a *mesa de trabajo* with Hidrocapital". However, MTA members were barely involved in other community-based associations (only 2 MTA members were members of the CTU as well) as well as scarcely informed about the activities of the other CBOs, while criticisms of these possible collaborations also apply. Some leaders highlighted the uselessness of waiting for the support of the municipality or water company, while they pointed out that the MTA should instead help community members to organise themselves and work together to solve service-related problems – in an informal, locally-based manner:

> it is impossible to obtain something [from the municipality or Hidrocapital]... it is not worth trying because there is no response nor results, it is a waste of time ... With the MTA, we are going to meet up and go to speak with a part of the community who is getting sufficient water, in order to ask them to share water, if there is the possibility ... if we ask through the MTA, they know that we have more knowledge; we know how to speak with people (MTA leader *terraza* 8).

Effectiveness in responding to civil society concerns

Responsiveness to community priorities in water delivery is in theory guaranteed by the fact that service improvements are jointly elaborated by the water company and the community, represented by MTA leaders, based on the household and water census, and on the cooperative design

7 As presented in the previous chapter, the water committee in HeC benefited from the direct support of the mayor who accompanied MTA leaders to the meetings with Hidrocapital officials and with the community, and who contributed financially to the pipeline construction through a disbursement from the municipal budget. As a result, the mayor had always been personally involved in and informed about the MTA physical results in the construction of pipelines and progress in the settlement upgrading.

of the water project. But, in practice, empirical evidence from the field shows how responsiveness is heavily affected by the capacity of the collaboration between MTAs, the water company and the municipality to produce real improvements in the water network, as well as effective communication between the technical water committees and the public institutions.

So, in the *Hacienda el Carmen*, local governance arrangements seem to have addressed community concerns for water delivery (Q1). The MTA in this settlement was created with the explicit purpose of reducing time and prices in getting water, by designing a water network that should allow dwellers not to rely just on tanker trucks or illegal connections and to avoid, in turn, related problems. After the water project, community members experienced many improvements. First, they were actually no longer forced to queue for water from the tanker trucks. Second, they also experienced a reduction in social conflicts and tensions linked to the purchase of full tankers by people closer to the main settlements roads, leaving other residents waiting for up to 4-5 days for the following trucks[8]. Third, the water project had dramatically reduced the cost of water: a resident of the *Colinas* sector – the only HeC's sector which was paying for water at the time of investigation – reckoned that he was paying less than 1/30 of his previous monthly disbursement for water. Fourth, community residents felt comfortable with the new system because they started to know the exact timeframe of water distribution and they were sure that their connection was functioning one day out of two. Fifth, residents also felt more confident about the water quality, given that it came from the main Hidrocapital distribution system[9]. In addition, Hidrocapital provided all households with a germicide powder to add to water stored in drums in the days when the network was not operational.

On the contrary, effectiveness of local governance arrangements in *PR2* in responding to community concerns appeared rather limited. Community dwellers sought in particular to avoid queues in getting water from the few illegal water points, to provide the whole settlement with the same access to water, to pay lower prices for water connections, to improve the quality of the illegal connections, and to provide better

8 From the survey conducted with community residents in the sectors *Araguaney Amarillo* and *Colinas*, it emerged that – out of the 23 people interviewed – 19 of them previously relied on tanker trucks to get water, while just 4 of them were using illegal connections.

9 A person living in the sector *AAm* affirmed that "before it was not hygienic because there were leeches in the drums and, in addition, we didn't know where the water came from". Actually, the few people who were still getting water from illegal connections expressed their fear about water quality because they didn't know whether their connections were watertight and whether they were getting water that was already treated.

information about the changes and developments of the water projects[10]. Even after Hidrocapital's intervention, access conditions for drinking water were mixed in the different terraces of the settlement and even within the same terrace, principally in respect to quantity and pressure; most illegal connections were still built with materials bought by residents and without specific technical support; information on how the Hidrocapital intervention was organised in *terraza* 11 and how drinking water should be provided to the whole settlement was still lacking. In addition, the fact that most community dwellers were no longer queuing at the water points was much more related to the development of illegal connections and informal arrangements by residents rather than to the MTA commitment. Nonetheless, interventions realised with the support of the MTA, in terraces 10 and 11, proved that the commitment of this organisation could considerably reduce the price of illegal water connections.

Communication amongst local stakeholders also plays a fundamental role in effectively responding to local concerns. Actors responsible for the delivery of drinking water discover a community's necessities in different ways. MTA leaders recognise community needs in accessing water because, first, they often face the same problems themselves and, second, the community directly advises them[11]:

> We know the problems [of the community] because we live in the same community, we are a part of the community and the problems that the community holds are the same problems that we face (leader in the *Araguaney Amarillo* sector, *HeC*).
>
> People know that I am the head of water and so, if they have a problem, they come to me, and if they don't find me, they send me a text message. If we can find a solution within the community, we do it with the help of [a community member who collaborated at the construction of the network], but if the work is much bigger, I call Hidrocapital who sends us a contractor (leader in the *Colinas* sector, *HeC*).
>
> They [community residents] ask me what is happening; I answer 'look, we are waiting for the project and we are also waiting for the houses, to see if it is possible to solve them together' … there are meetings or when the people see someone in the street, or if they come to my house to ask what it is happening, what they [Hidrocapital officials] are doing … in this area, they come and they ask me …they know that I am in the MTA and they know the solution I can give them' (MTA leader in *terraza* 11, *PR2*).

---

10 Information collected through interviews with community residents in *terraza* 11, through the focus group and face-to-face interviews with MTA leaders.

11 The water survey (*censo del agua*) carried out by MTA members provided specific information about the conditions of each household and allowed settlement dwellers to know who was dealing with the water delivery problems.

Nevertheless, when we look at the way in which public institutions are made aware of water related problems, the situation changes once more in the two settlements. In Paz Castillo, Hidrocapital representatives and the mayor admitted being familiar with the concerns of the community because of the complaints they received directly from the same community at the Hidrocapital offices or at the municipality, but also because of street manifestations and protests against water shortages. Together with municipal engineers, they confirmed that there was regular communication between Hidrocapital and the municipality to keep informed about water related problems in low-income settlements[12]. In the municipality of Cristóbal Rojas, on the contrary, municipal public servants admitted to gaining some awareness of the community's problems in the same way, but that the relationships with MTAs were often tense and that the technical water committees were keener to directly go to the water company. The director of the "Urban and Rural Planning" department clearly stated that:

> Sometimes the community comes to the municipality or sometimes it goes directly to Hidrocapital … I am engaged with the community – and the community knows me – but I have never been called by an MTA … if I want to know something about the MTA, I ask Hidrocapital, this is how I get informed, but it is not the community that is telling it to me … I have contacts with many people there, with the *Concejo Comunal* and the *CTU* … but not with the MTA and, in my mind, this is an indicator of their capacity to be effective or not; it is a good instrument to get things done but, are they doing it well? I don't know! They mistrust the municipality and so on, well, come here and speak with us!

6.2. Micro-inequalities leading to spatial fragmentation

Governance mechanisms able to guarantee equity in water provisions at the community level need to consider both distributional and procedural equity. Concerning distributional equity, fair access to water supply is analysed according to, first, the factual distribution of water service improvements in the settlements; second, the extent to which improvements produced by joint collaboration among the MTA, the municipality and the water company meet community values for water, here expressed as the need for water improvements and willingness to pay for the service; and finally the supply aspect of service access, which include the extent of in kind contributions from dwellers, the service costs structure and the payment collection management. These features are investigated with respect to their capacity to contribute to both individual and community access to services.

---

12 Hidrocapital was organising a meeting once a month with public servants from the different Tuy municipalities.

First of all, if we look to the water service improvements obtained by the MTAs by lobbying and collaborating with public institutions, an equitable distribution of such results does not systematically occur. Precise comparisons between *Hacienda el Carmen* and *Paso Real 2000* are hard to do given that the respective MTAs were at two very different stages in the process of improving their water provisions; but it clearly emerges from the fieldwork that both settlements (although most significantly the *PR2*) were experiencing different conditions accessing water provisions within their boundaries. In all areas where the water company and the technical water committee gained service improvements, residents accessed water unevenly since the timeframe and the pressure varied according to their relative height and distance from the main pipelines. In *terraza* 11 of the *Paso Real 2000* settlement, besides this kind of unequal service access linked to the morphology of the terrace, some households also faced inequitable access to water since they were not provided with any form of piped connection but obliged to rely on neighbours' willingness to share their personal water point.

Second, concerning the community values attached to water, residents' needs and desires for service improvements obviously embrace good water distribution, including the quality, quantity, pressure, frequency and reliability of water supply, but also less technical concerns related to an equitable distribution of water throughout the whole settlement, or to a renewed culture of water through which to learn not to waste it and to accept reduced provisions. The extent to which the most technical and distributional values are met by the piped networked project clearly depends on the results obtained through the water project, while the extent to which the most cultural and social values can be realised may clearly represent an interesting research subject for scholars in the future.

Third, in relation to the financial organisation of water supply, when the local governance arrangements have produced service upgrades (in the *Hacienda el Carmen* and *terraza* 11 of *Paso Real 2000*) the water prices have been dramatically reduced, thanks to Hidrocapital's technical and economic support, the (although very limited) municipal financial support and the removal of informal connections between neighbourhoods (which entailed very high water prices for the residents using them). The application of a minimum tariff in the *Colinas* sector – which is proposed in the other sectors too – had also contributed to a cut in the prices paid by each household for water and encouraged people to pay for the service. Willingness to pay for water might be taken for granted, at least on a theoretical level, in both settlements. This relies on the dwellers' awareness that they were paying more for illegal connections or purchases from water trucks, but also on the feeling of being recognised as citizens in their settlement, through the institutions' commitment to providing them with water, and the possibility of receiving an institutional response to their water shortage. Nevertheless, the community members' reactions to their individual meters and to the payment collection organisation will only be assessed after these individual meters are installed and the management of payment collection is made operational.

Regarding procedural equity with respect to community involvement in local governance, the decision-making processes developed around the technical water committees seem to have largely relied on agreements between the MTA and the water company, with limited involvement from community members and an almost totally absence of the municipality. A partial exception is represented by the PR2 settlement, since the whole community has been involved in the few assemblies organised to identify MTA members and present water payment systems (see also § 6.3.2 on participatory mechanisms at the settlement level).

Before moving on, a remark has to be made with respect to the gender dimension included by UN-HABITAT in the UGI indicators related to the equity principle (UGI indicators No. 10 and No. 11). Contrary to these indicators, we do not investigate gender equity in the governance mechanisms built around the MTA at the settlement level. From the first site visits, in effect, it was clear that MTAs and condominiums in the *Hacienda el Carmen* settlement were almost exclusively run by women, while in *Paso Real 2000*, men and women belonging to the MTA were present in an equal number. In the same vein, according to UN-HABITAT (2006a: 83), in the peripheral settlements in metropolitan Caracas "these water committees also provided more scope for women's involvement – they represented three quarters of those who took part and also many leaders. Some have also begun to work on the community issues". The first main reason of this strong female participation has been identify by MTA leaders in the greater time availability of women with respect to men.

Distributional equity: access to basic services

Distribution of water improvements

Despite the improvements experienced by the *Hacienda el Carmen* in the effective delivery of drinking water, community residents were accessing water with slightly different conditions, revealing a sort of spatial inequality. First of all, in the sectors *Araguaney Amarillo* and *Colinas*, all households were connected to the same piped network, but they received water with a different pressure and timeframe according to their relative height and distance from the main pipes. So, people living in the lower parts of the settlement and close to the origin of the settlement's main pipe received water all day with good pressure, while those remaining (the majority of interviewed dwellers, 19 out of 23) received water once every two days with variable pressure. In addition, the *Nueva Republica* sector was lacking water because of the recent illegal connections that a new informal settlement (*Los Muños*) had opened from their pipelines. The MTA leader in this sector reported that many residents of this area confirmed they were much better off before the Hidrocapital project, when they relied on their informally and illegally made connections.

In the PR2 settlement, the differences between individual and community access to water are much deeper. Some areas were actually provided with quite a good service, night and day, except for one day a week (e.g. terraces 2 and 7); some areas accessed water through illegal connections but only 6-7 hours per day, 6 days per week, but in a small quantity and with low pressure, according to their relative height and distance from the main pipes; other areas were totally lacking any form of access to water (e.g. a part of terrace 10 and half of terrace 8). Furthermore, some households had direct (illegal) access to Hidrocapital water pipes, while others accessed water through plastic hosepipes attached to neighbours' connections. *Terraza* 11 was probably experiencing the worst inequality in drinking water provisions among community dwellers. Residents, in fact, were accessing water through (see also map 10):

- pipelines built via the Hidrocapital intervention in 2005 (51 households),
- pipelines built via the Hidrocapital intervention in 2005, but experiencing significant lower pressure in the pipelines (5 households),
- illegal pipelines from the water company's main pipe "below" providing water 6-7 hours per day, except for one day per week (19 households),
- illegal pipelines from the water company's main pipe "above" providing water night and day, except for one day a week (6 households),
- plastic hosepipes attached to neighbours' piped connections (16 households),
- no access to water (at least 2 households).

Meeting community's values for water

Needs for water improvements

Improvements in water distribution in the *HeC* seem to match residents' wishes, as we have already shown in the section concerning effectiveness in responding to civil society concerns (§ 6.1.2). In most cases, community members declared themselves to be much better off with the new piped water network (Q11). Only a few people expressed their desire for further improvements, such as accessing water seven days a week with higher pressure (through larger pipelines) and educating people not to waste it; some other residents of the *Colinas* sector expressed their desire that everybody pays for his own water usage. There were also some people arguing that the new water system and timeframe was reasonable "because the service has to be provided to other communities as well" and "because in this way people can understand that water is important".

In the *Paso Real 2000* settlement, where the Hidrocapital project had not been realised, the values that the residents assigned to the water services were essentially represented by the necessities that they still required for an improved water service. It is interesting to notice that, in *terraza* 11, the desired improvements did not vary significantly among the residents accessing water through Hidrocapital intervention and those who were still relying on the previous illegal connections (Q11, see Annex 2), confirming the limited impact of the water improvements obtained through the hydrological company in this area. In particular, community dwellers in this terrace required three main changes. First, they would have seen improvements in the quality of the pipelines since they did not feel very confident with the informal network, because of multiple connections to the same pipelines and because of the "provisional" character of most of them. Second, many dwellers thought that the project should have been extended to the entire settlement, not only for a sense of community justice and equity with respect to service access, but also because inequitable access conditions harmed all dwellers: one resident affirmed indeed that "below, people offer water to neighbours and as a result we lack water; as a consequence, those who paid and worked for the project are not accessing water". Finally, some other residents expressed their desire to improve water services together with the *CTU* project of *substitucion de rancho por vivienda*, in order to address the problem of urban services jointly and realise an all-inclusive development plan in the settlement.

Willingness to pay

All community residents stated they were willing to pay for their water consumption (Q10a)[13]. In general, people agree with paying for water because they consider it an essential resource and they link payment with being able to complain about the network's inefficiencies (see Annex 2). But on the ground, the situation can be different. In the *Colinas* sector (HeC settlement), for example, only 2/3 of people confirmed that they were paying for water, whilst the MTA leader of the same sector

13 Just one person in the *HeC* settlement disagreed with water payment: he actually wanted an individual meter and a direct receipt from Hidrocapital for his own consumption.

reckoned that just 1/3 of residents was paying. According to this water committee leader, many people were justifying their refusal to pay by arguing that "Chavez said that we don't have to pay for water" or that "the state government said that we have one year of grace" in which they are exempt from payment. In *Paso Real 2000*, on the contrary, we could not find any empirical evidence related to residents' willingness to pay, as residents built their illegal connections by themselves and the whole settlement had never had to face any official payment for water. The only exception was the financial contribution that dwellers in the 11th terrace had to provide for the network's construction: in this case, according to an MTA leader in *terraza* 11, not all the residents paid their connection fee (of 15,000 *bolivares*).

Equity in water supply

Equitable access to water services is significantly connected not only to the way in which community residents contributed to the piped network construction to ensure its widespread completion throughout the entire settlement, but also to the financial architecture to guaranteeing service affordability. Where improvements to the water network have been realised, as in the *Hacienda el Carmen*, they have rested, first, on different types of "in kind" contributions by community residents and, second, on a cost structure and payment collection system allowing low-income dwellers to pay for the service.

In kind contributions

As Cariola and Lacabana (2005a) claimed, governance arrangements for better water provisions in Venezuela are based on the involvement of and contributions from the MTA and community members, from the beginning to the end of the water delivery process. This means that residents (such as those in the two sectors of the HeC), volunteer in different ways to improve the water delivery system. From one side, through the availability of MTA leaders and their unpaid work in participating in meetings with the water company officials and in following up these meetings within the community; in organising and working out the water census, the settlement map and the water distribution project; in receiving and hosting engineers from the hydrological enterprise during the pipeline installations (as emerged from the focus group with MTA leaders). From the other, through the availability of community dwellers participating in the census and organisational meetings, and the voluntary work – especially from men – in the construction of the piped network. According to the ex-community coordinator of the Losada Ocumarito System), this community involvement relies upon a specific idea of service supply, in which communities play a role as important as that of the water company:

> The MTA is a community organisation that is set up with the purpose to solve the problem of water *with* the community ... according to the necessity to make people manage and self-manage their own reality. ... At the beginning, Hidrocapital was a totally technical enterprise, dealing with water collection, treatment, distribution and commercialisation; then, we changed it because we incorporated the humanisation of this enterprise, because this enterprise was exclusively technical ... now it is technical and human ... this is the reason why the MTAs were constituted: for the government, there are viable solutions. ... The idea was to support the community in doing this [participating at the delivery process] as a brother, not to do it in its place; the idea is that, this time, it is a question of aid and solidarity.

The water committees' leaders in the HeC have operated in this sense: the inclusion of community residents (members of the community or of the MTA) in the construction of the piped network has meant not only a clear responsiveness to the community's demands, but also a greater capacity to self-manage the service by the community. MTA leaders acknowledge this new shared and collaborative approach to settlement upgrade as a way to directly support communities and involve them in the service delivery process:

> The purpose of the MTA is to obtain the service, the project for the community; to know technical things about the construction and the maintenance of the pipeline network ... the MTA is – as Hidrocapital says – an instrument of revolution: where there are communities who do not have the possibility to make a project, through the MTA, they can obtain the project and organise themselves (leader in the *Araguaney Amarillo* sector, *HeC*).
>
> ... we have to be organised to manage the work [the drinking water network], it is more serious; the MTA of the *Hacienda el Carmen* is recognised by Hidrocapital, it is like having more respect, more seriousness; it is valued because it represents the *HeC* (leader in the *Colinas* sector, *HeC*).

On the ground, nevertheless, this view is not fully shared and perceived. Community dwellers in the *Hacienda el Carmen* revealed quite a different opinion with respect to the MTA's role in organising service provisions. First of all, only 6 residents out of the 23 interviewed had any idea of the MTA's purpose (Q7), while the remaining generally knew the person in charge for water-related questions in their sector but ignored the existence of the MTA – especially in the *Araguaney Amarillo* sector (Q2a). In addition, those dwellers who knew the MTA or the condominium recognised the very practical role of these organisations, confirming that their purpose was to technically provide the service to the community and to control payments for water (see Annex 2 for detailed answers). Still, during the interviews, they never made any reference to the technical water committee's role in involving the whole community in the decisions and engagements over the service provision process or in transmitting their needs and complaints to the hydrological enterprise. Contributions from community members to the improvement of the water delivery were thus identified just with their operational contribution to the network construction rather than with their contribution to a shared and inclusive development process, as underlined by MTA leaders and Hidrocapital officials. This vision also applies in the case of an absent or ineffective water project, as in the *Paso Real 2000* settlement. Here, both MTA leaders and residents identify the purpose of the technical water board exclusively with its material task to provide drinking water to the whole settlement, without any specific reference to their role in including the entire community (see also Q7 and Annex 2):

> The purpose of the MTA is to obtain the service because it is what we need ... the service has to be for everybody ... The function we have as the MTA is trying to intermediate between the water company – which is going to give us the service – and the community, this is our job, but it not to go to see if a pipeline is repaired or that there are people fighting for water ... this is not my task, ok, we do it just because we like a nice atmosphere in the settlement ... a leader can mediate ... the MTA makes a commitment to provide just the water company with the information necessary to the company, the water company ask you for a community survey about the houses, then they ask you to do a social study to see your possibility to pay for the service according to your level of life ... then you have to bring all these things to the water company with *croquis* and the utilisation of the land ... we brought all the requested things, whilst the only thing missing is the topographic survey (focus group with *Paso Real 2000* MTA leaders).

Water cost structure and payment collection

The financial viability of the improved water network relies on the affordability of the costs necessary to connect households to the piped network and also on the effectiveness of payment collection. Concerning connection costs, these were much cheaper than the average household disbursement for water tanker trucks or the previous illegal connections. In the HeC settlement, for instance, illegal connections to someone else's pipeline could cost up to 100,000 *bolivares* (46.62 US$); while in *Paso Real 2000* the highest connection cost was paid in terraces 2 and 7 (80,000 *bolivares*)[14]. With the MTA-Hidrocapital project, connections to neighbourhood water networks were free and dwellers only paid the cost of the brackets purchased in order to be connected to the main pipes (10,000 *bolivares* on average in the HeC, and 15,000 *bolivares* in

[14] One resident in the 11th terrace affirmed that he was asked to pay 180,000 *bolivares* to get an illegal water connection.

terrace 11 of *Paso Real 2000*). Construction and connection costs, and their sharing among the water company, the community and – to a lesser extent – the respective municipalities, are presented in table 6.1.

Table 6.1. Distribution of construction and connection costs[15]

	Hacienda el Carmen	*Paso Real 2000*
Water company	all the settlement: costs related to the provision of the main pipelines (6 to 8 inches), of technical machinery and professionals (engineers)	*terraza* 11 (half): costs related to the provision of the main pipelines, of technical machinery and professionals (engineers)
Community	cost of bracket to make the connection to the main pipe and of the secondary hosepipes (generally ¾ to 1 inch) from the main pipe to the house: . sector *Araguaney Amarillo*: 12,000 *bolivares* per household . sector *Colinas*: 10,000 *bolivares* per household	– cost of bracket to make the connection to the main pipe and part of the secondary hosepipes from the main pipe to the house: . *terraza* 11: 15,000 *bolivares* per household – cost for illegal connection: . *terraza* 10: 29,000 *bolivares* per household . *terrazas* 2 and 7: 80,000 *bolivares* per household . *terraza* 8: not available
Municipality	sectors *Araguaney Amarillo*, *Colinas* and *El Esfuerzo*: financial contribution by the Paz Castillo Municipality to the purchase of secondary hosepipes, not covered by the community[15].	*terraza* 10 and 11: contribution through the offer of the secondary pipelines (2 inches)

Concerning payment collection, if in *Paso Real 2000* water was not yet metered and billed, in the *Hacienda el Carmen* settlement the *Colinas* sector had been paying for the water service for a year, according to the social tariff, while other sectors were about to begin water payments. Although there was a broad willingness to pay for water among residents, technical water committee leaders argued that, eventually, many people were refusing or were likely to refuse to pay. In the *Nueva Republica* sector, for instance, residents were threatening Hidrocapital and the person responsible for the sector with their refusal to pay for water if no solution to their current water shortage was found (a new informal settlement, next to theirs – *Los Munos* – was building illegal connections on their pipelines, thus reducing their water pressure). The leader of the *El Esfuerzo* sector said that many people were only there at the weekend and so, it would have been particularly difficult to collect water payments

15 In the *Araguaney Amarillo* sector the contribution added up to 1,686,500 *bolivares* (786.39 US$), while in the *Colinas* sector, it amounted to 600,000 *bolivares* (279.77 US$). For the *El Esfuerzo* sector there are no figures available

from them. In the *Araguaney Amarillo* sector – but also in others – people asked for an individual meter in each house and a Hidrocapital receipt for their water consumption (thus promoting greater transparency in payment collection)[16]. Finally, as we have already noted, just one third of dwellers in the *Colinas* area was actually paying for the service.

But these are not the only concerns associated with the collection of water fees. With respect to the management of water payment collection, the technical water committee leaders confirmed their availability for gathering these payments but, at the same time, admitted their concerns about the possible suspiciousness regarding community leaders' honesty in managing residents' money. Indeed, MTA leaders confessed that they face people's distrust when dealing with money. Hidrocapital officials and MTA leaders seemed to be well aware of this situation and, to face the water payment problems they were trying – amongst other initiatives[17] – to set up a cooperative. The future cooperative in the HeC was supposed to carry a twofold practical function: collecting water payments and maintaining the piped network. Together with the creation of a cooperative, the organisational structure of the condominiums would have disappeared and Hidrocapital would have provided each household with an individual meter[18].

The same MTA participants, together with other community members who took part in the pipeline construction and thus acquired the technical knowledge to maintain the network, should compose the HeC cooperative. The underlying idea of this organisation was not only to strengthen community commitment to water payment, but also to support local forms of social economy (Cariola and Lacabana, 2003). According to the community coordinator of the Losada Ocumarito System, "the cooperative is within the MTA because the water payment should be realised within the same community, and because the community should be aware of the water payment. ... The commitment of Hidrocapital is that the water has to be paid for but also that community could gain from water payment collection through the cooperative".

16 An MTA leader in the *Araguaney Amarillo* sector said that many people claimed she had bought her house with the money collected for the water project.

17 Amongst other initiatives, the MTA in the HeC settlement – with the support of Hidrocapital – is organising meetings to explain the functioning of this payment. In addition, the media (television, radio, paper advertisements, etc.) actively promotes the fact that the state wants to provide all citizens with water and how important it is for the population to be co-responsible for the service through a "just payment" of water.

18 Interview with the community coordinator of the Losada Ocumarito System. In particular, according to the ex-community coordinator of the Losada Ocumarito System) "the individual meter has the objective to make people aware of their consumption ... the water meter is more a social meter than an economic one".

From the point of view of the MTA leaders, the cooperative would organise payment collection by delivering individual receipts to each household and directly collecting the payments. The advantages derived from the setting up of a cooperative were identified by the MTA in the income revenues for cooperative members stemming from this activity (who should receive a percentage of the total fees collected), the sharing of payment collection among these members according to their time, availability and, above all, the possibility of avoiding personal threats from those neighbours unwilling to pay. With respect to the latter, MTA members were particularly supportive of the cooperative initiative insofar as they could request other cooperative members to collect payments from insolvent households in their sector. It is nevertheless clear that, despite this last expedient, it is still possible that a cooperative's management could generate social conflict and vengeances within the settlement.

Procedural equity (decision-making process)

The shared decision-making initiative is at the foundation of the participative democracy reforms that have been taking place in Venezuela. Water provisions have always played a fundamental role in the construction of this political project:

> How do you construct local power? Because this is a local power of government; the MTAs and the CCHs [*Consejos Comunitarios de Hidrocapital*] are an application of the government's participation in decision-making, in order to realise the project and to know if the project is endorsed by a specific community ... to verify if the leaders are false. For the state to invest in community water projects, it is a necessary and important discussion. ... All the projects for water pass through discussion with the people ... the CCHs were an example for the current Local Public Planning Councils (ex-community coordinator of the *Losada Ocumarito* System).

On the ground, though, although MTA leaders and Hidrocapital officials tend to explain that the water project is based on essential and decisive collaboration with the residents, it seems that community members did not had a relevant say in most of the important decisions concerning improvements to the piped network. In the *Hacienda el Carmen*, for instance, the delimitation of condominiums together with the design and organisation of the piped water network have been determined by Hidrocapital and the technical water committee, without any particular contribution from the community. The involvement of community members in the designation of the people in charge of the condominium[19]

[19] In the *Araguaney Amarillo* sector, four groups of people belonging to the four main streets of this area have been elected responsible for the *Araguaney Amarillo* condominiums, while in the *Colinas* sector, nine people have been elected responsible for the condominiums.

also appeared to have been very limited: just one person[20] among the 23 dwellers interviewed remembered this election process (Q8b).

If we look now to the payment system's organisation, the situation is very similar. In both communities, dwellers were asked about their preference for an individual or collective meter and the communication method for water service bills (during the water census in the *Hacienda el Carmen*, and during a community assembly, open to all residents, to present the two different payment systems in *Paso Real 2000*). Nevertheless, the final decision in the HeC was made by the water committee and the water company, while there was no definitive choice in PR2, despite the residents' preference being for payment through the condominium[21].

Finally, in both settlements, there are no official and regular meetings between the technical water committees and the residents. In general, in the areas where the water project was realised, dwellers usually know the people in charge of water-related problems (Q2a) and they know that they can urge them to find a solution to water shortage issues (Q4 and Q5). But, most of the residents claimed to have never spoken and/or that they would not be willing to speak with the water committee leaders in the case of a service shortage, essentially because they maintain that, if there is a problem with the water pipelines, they solve it by themselves or that MTA leaders also have the same problems, making it useless to inform about the issue (Q4 and Q5).

6.3. Efficient informal participation

If representative democracy at the settlement level can be assessed by looking at the procedures through which MTA leaders were elected, participative democracy in the technical water committees is a much more complex phenomenon.

20 This person is nonetheless close friend with one of the *Araguaney Amarillo* MTA leaders who was present at the moment of the interview.

21 "Yes, we held a meeting to speak about the water payment in July of last year (2005) … the assembly was to communicate discussion to the community and the company about how to pay for the service and we proposed payment through the condominiums … Hidrocapital gives the option to the community to choose between the community or paying individually … the decision about the meter was not taken by the MTA but within the community, we asked them if they were more interested in paying individually – where it is necessary to install the meter, to pay for the meter, to pay for the payment receipt, to go to the bank to pay the quota – or payment through the condominium which is less expensive … someone wanted to be independent because some people use more water than others, and he did not want to pay a single tariff … but as usual the majority prevails and the majority wanted the payment through the condominium … the important thing was that the community had to choose how to pay the service" (MTA leader in the *terraza* 11, PR2).

> The MTA is a tool of revolution, because it is born as a direct consequence of the ongoing changes in the country … the MTA is useful to make it clear that the water problem is not just a problem for the hydrologic enterprise but it is a problem for everyone; the MTA is committed to working with the community and the community organises itself in assemblies to find the best solution to the problem of drinking water and sanitation; the protagonist participation is central (*participacion protagonica*) because they [people] control the work and they directly realise the project with the hydrologic enterprise … it is a joint project between the community and Hidrocapital (coordinator of community participation for the Losada Ocumarito System).

As previously expressed, the technical water committee constitutes the essential link of the improved water delivery chain, ensuring participation both downward and upward. So, from a downward perspective, they represent the formal organisation through which low-income communities can claim for the provision of water services and channel their participation in the improvements' decisions and realisations. From the upward perspective they are in charge of the relationships, contacts and communications with the water company and the municipality. In the first case (downward), an estimation of the extent of "participation in practice" (Tunstall, 2002), at the local level, may be captured through the level of knowledge of community members with respect to the existence and functioning of MTAs and/or condominiums[22], the degree of community participation in settlement meetings, and the extent of residents' involvement in the construction of the piped network. In the second case (upward), participation might be associated with the quality of the relationships developed within the collaborations amongst the municipality, the water company and the technical water committees. Both formal and informal arrangements emerge when stakeholders participate, communicate and collaborate at all these different levels.

An accurate evaluation of the differences between the two settlements in the extent of stakeholders' participation mechanisms and processes allows us to outline a possible trend in participation on the ground. Looking at the two MTAs, in the HeC settlement, few residents know the water technical committee organisation (mainly thanks to the door-to-door payment collection); community participation appears limited to specific circumstances (meetings for and collaboration in the construction of piped network); collaboration mechanisms among the main local

22 According to Arnstein (1969), the awareness of the existence of a participatory tool is just a minor form of participation ("*informing*") but it is nevertheless a first step to real citizens' participation.

stakeholders (but especially between the water committee and the water company) appear characterised by a strong and continuous participation and are principally carried out in an informal and flexible way; and the election of MTA leaders has also occurred in a rather informal way, based on the previous commitment of community leaders to other CBOs. On the contrary, in the PR2 settlement, beside the prevalence of informal communication between the community and MTA members, and the absence of a comprehensive piped water network, many residents know of the MTA and the process through which it has been set up (mainly through the initial general public assembly and communication within the areas); a more formal and open procedure involved the whole community in the selection of the water committee leaders; and the very limited interactions among the local actors are characterised by their formal appearance (the MTA leaders' scarce direct involvement with municipal and water company officials and formal requests of assistance to local institutions, except in the case of Hidrocapital's intervention in terrace 11, which followed very insistent and direct pressure from the MTA leader).

Consequently, a positive relationship between improvements in water provisions for low-income dwellers, and the kind as well as the extent of relationships among local stakeholders on the ground can be deduced. Service upgrades appear to be more an outcome, first, of the informal, flexible and context-specific contacts and collaboration between MTA leaders and municipal and water company officials, rather than of the precise adoption of official communication channels. Second, improvements in water delivery are much more linked to a limited community participation in specific activities, rather than to an open, widespread, inclusive and more formal communication with and involvement of community members in the functioning of the water committee (leaders' elections, water committee's role, etc.).

If the quality of the relationships among local stakeholders is dependent on their informality and suppleness, the extent of these collaborations confirms that Hidrocapital and the MTAs are far more present and involved in these processes than the whole community or the municipality. Actors on the ground (according to the interviews run and the scoring exercise presented in tables 6.2 and 6.3), first, confer to the municipality the lowest contribution for the effective delivery of drinking water; second, although they attribute an important role to community members, in practice, community involvement seems often to be quite limited (tables 6.2 and 6.3 show that the contribution of residents is judged more important in PR2, where dwellers solve their water shortage by themselves, rather than in the HeC settlement, where the MTA obtained, managed and realised the water project).

Table 6.2. Relative importance of local stakeholders in the *Hacienda el Carmen*

	1 = very low	2	3	4	5 = very high
MTA leaders					
community members				••	•••
MTA			•		••••
Municipality		••		••	•
Hidrocapital		•	•	•	••
Municipal public servants					
community members			•		••
MTA			•	•	•
Municipality		•		••	
Hidrocapital			••		•
State-owned water company officials					
community members				•	••
MTA					•••
Municipality		•	•	•	
Hidrocapital				••	•
TOTAL					
community members			•	•••	•••••••
MTA			••	•	••••••••
Municipality		••••	•	•••••	•
Hidrocapital		•	•••	•••	••••

Table 6.3. Relative importance of local stakeholders in *Paso Real 2000*

	1 = very low	2	3	4	5 = very high
MTA leaders					
community members					•••
MTA				•	••
Municipality	•	•	•		
Hidrocapital			•		••
Municipal public servants					
community members			•		•
MTA		•	•		
Municipality				••	
Hidrocapital			•		•
State-owned water company officials					
community members		•			••
MTA				•	••
Municipality		••	•		
Hidrocapital				••	•
TOTAL					
community members			•	•	••••••
MTA		•	•	•	••••
Municipality	•	•••	••	••	
Hidrocapital			••	••	••••

Participative democracy

Dwellers' knowledge about MTAs and condominiums

No general trends are identifiable with respect to the dwellers' knowledge of the existence and functioning of MTAs and/or condominiums. About half of the interviewed residents were aware of the existence of a *mesa tecnica de agua* or *condominio de agua* (Q2a), through a great variety of information channels (Q2b). In the *Colinas* (HeC) sector, for example, where almost all residents knew the MTA or the condominium, community residents were aware of these organisational bodies mainly because of the door-to-door payment collection operated by the MTA leader and because of the meetings held in that sector to present and organise water payments. In *terrace* 11, in the *Paso Real 2000* settlement, residents affirmed knowing the MTA through the initial public assembly organised in 2003[23], through direct and personal communication from the MTA leaders and through the water survey that they ran in each household.

Participation in MTA meetings

The participation of community residents in the technical water committee meetings organised by MTA leaders is one of the two ways (together with in kind contributions to the network construction) in which these two actors are jointly involved in the water delivery process in practice (Q9). Meetings between the MTA leaders and community members are not generally organised on a regular basis, but according to the necessity to involve residents in specific (often practical and technical) matters (Q8a). In the HeC, for instance, the MTA have organised community assemblies for the technical organisation of the network's construction, the payment sharing to buy the hosepipes (in the *Colinas* and *Araguaney Amarillo* sectors) and the MTA's communications concerning the water payment system (only the *Colinas* sector). A small number of meetings in Paso Real 2000 were organised to inform residents of the possibility to constitute a technical water committee, on the few MTA progresses in the execution of the service, and on the realisation of a partial project in *terrace* 11. If there is no progress with the water project (PR2)[24] or the

---

23 A public assembly was organised in *Paso Real 2000* to present the possibility of organising a technical water committee, and of participating in this association, on a voluntary basis. Invitation to this assembly was open to the whole community, while resident participation was considered quite good by MTA leaders, although only about 200 people (from a population of about 800 households / 3,000 people in the whole settlement) were present.

24 "We want to organise assemblies with the residents, as for the *Comité de Tierra Urbana*, we have to! But the problem is that we are waiting for the topographic survey,

new network is finalised (HeC), both residents and MTA leaders admit that no assemblies are organised, confirming that they have a much more technical rather than consultative or debating character (see § 6.4.1).

Information concerning meetings was generally provided in an informal way. According to MTA leaders in the *Hacienda el Carmen* settlement, for instance, they have personally informed the community about the meetings or have sometimes hung up small paper notices around the settlement. As a result, most of the residents were aware of assemblies concerning the water service through the information provided by MTA leaders but also through conversations with their neighbours (Q8a)[25]. Nevertheless, community participation in these meetings appears quite limited[26]: MTA leaders, in particular, recognised strong community participation with regards to the piped connection payments and the labour contribution for the realisation of the network but, at the same time, they stressed that generally "the community is apathetic" and not interested in MTA activities. The leader of the *Araguaney Amarillo* sector also confirmed "there are many people thinking that, if someone is going to obtain the service, they [people] are going to obtain it just the same, without doing anything".

Community involvement in the network construction

Without a doubt, contributing to the concrete realisation of the water network is at the core of the community residents' participation in MTA activities. In all analysed cases, where water improvements have been realised through a piped network, all dwellers have at least personally provided their households with the hosepipes, and the large majority have also directly participated in the construction of the main pipes with Hidrocapital engineers (see also § 6.2).

---

we don't have it and how are we going to have an assembly if we don't have anything to say? If we don't have any information ... we have been waiting for 3 years (MTA leader terraces 2 and 7, PR2).

25 However, some community members complained about the scarce visibility and the incorrect information contained in the small paper notices. In our experience, we participated in a meeting of the *Consejo Comunitario* in the *Araguaney Amarillo* sector for which two different small paper notices had been hung up – providing different meeting locations.

26 Just 8 people, out of the 23 interviewed, affirmed participation in these assemblies (Q9). MTA leaders declared, that, usually, in the *Colinas* sector about 50/60 people attended the meetings (out of the more or less 450 people – 290 households – composing the sector), while in the *Araguaney Amarillo* sector, meetings were attended by 80/100 people out of about 1,700 people (695 households) within the whole sector.

Collaboration between the municipality, the water company and the technical water committees

There exists a two-fold connection between the quality of the relationships among the local actors and the water delivery improvements. If it seems quite obvious that progress in the piped project is positively related to good relationships among the local actors, these improvements appear also to be directly associated with informal and direct communication and collaboration mechanisms, rather than to more formal participation schemes. The cooperation between the technical water committees with Hidrocapital and the municipality clearly shows this.

In the *HeC* settlement, for instance, MTA leaders and Hidrocapital officials deem the relationships excellent, without any relevant exception. During the realisation of the water network (project definition and construction), they constantly worked together, making it easier to build up personal and reciprocal respect and familiarity. Moving on to the municipality, opinions among technical water committee members about the quality of their contact with the local government are not so homogeneous. In general, judgements vary accordingly to the level of economic contribution given by the municipality. They range from an enthusiastic description of municipal participation in the *Araguaney Amarillo* sector where the mayor was extremely present[27], to a "completely negative" judgement by a representative of the *Nueva Republica* sector that did not benefit from any economic contribution from the municipality. Opinions also include more nuanced judgment, showing nonetheless that the participation of the municipality in water improvements was pretty limited.

> The mayor was present at the beginning of the construction of the water network but then he was quite apathetic … he sent his sister to the water network inauguration. He came here again in the settlement just when the elections were approaching. … Nevertheless, when we needed MINFRA maps to organise the water project, the municipality gave us the plans … the mayor also put a great amount of the money in the power network project (an MTA leader of the *Colinas* sector).

The communication between the technical water committee leaders and local institutions came about through personal and informal contacts, rather than official information and communication channels. In the case of a technical problem with the pipelines, for instance, MTA leaders

27 In this sector, "the mayor has always been available to help us, regardless of the fact that we are occupying this land illegally", by economically contributing to the purchase of pipelines for the water network and covering almost all the costs of the power network, as the community's contributions were very limited (an MTA leader of the *Araguaney Amarillo* sector).

called Hidrocapital officials directly on their mobile phones to ask for a network repair. Nevertheless, this type of direct communication appears to function only with some MTA members. A water committee leader of the *Nueva Republica* sector admitted she asked Hidrocapital for a long time – by phone and in person – to find a solution to their water shortage, without getting a reply. The MTA leader of the sector *Colinas* confirmed that she invited community members to call Hidrocapital officials directly but, eventually, this more straight community participation did not work because of both the resident's reluctance, and the different behaviour of water company officials with respect to people they did not know personally. Good contact between MTA members and Hidrocapital officers depended ultimately very much on the individual and personal relationships between them.

With regards to the settlement of *Paso Real 2000*, joint participation between the MTA, Hidrocapital and the municipality seems to have been very limited when compared to the situation of the *HeC* settlement. According to an MTA leader in *terraza* 11, relationships between the technical water board and the water company were quite cordial, in so much as "when they [Hidrocapital representatives] receive us, they do it well, they are very kind, officials are very qualified, educated..." Nevertheless, there was a widespread disappointment with respect to the huge delays that the community was experiencing in getting an answer to their continuous requests for a piped water network:

> we have fulfilled all our duties, now we are waiting for them, but we don't know when they will do something ... they [Hidrocapital engineers] hide themselves behind the fact that they are not going to work if the state doesn't guarantee that it will cover the costs of the topographic survey, or the pipelines, or the workers ... resources are what stop all the works ... (focus group with PR2 MTA members)[28].

The lack of concrete responses make the technical water committee leaders deem contact with the municipality to be totally negative. Although they recognised the essential role of the municipality in lobbying the water company to improve water provisions, they also stressed they were short of any support from the local government:

---

[28] In reality, the reason for the stalemate of the water-networked project in the PR2 settlement was not so clear and there were inconsistencies between the MTA and Hidrocapital's expectations about the kind of intervention in this area. MTA leaders argued that the project was blocked because a water company engineer had not yet realised the topographic survey and, thus, the community could only wait for it. On the other side, Hidrocapital officials claimed that intervention in that settlement was not a priority since PR2 residents had quite a good service (interview with the Hidrocapital general manager of the Losada Ocumarito system and with the community coordinator of the Losada Ocumarito system).

> It is necessary that the community, the municipality and Hidrocapital work together to find a solution to this problem. The municipality should represent the link to let this project proceed faster but if we do not have the support, the help of the mayor or the municipality it is longer, I think that if she [the mayor] had taken care of this project a little more, this project would have already been finished (MTA leader *terraza 10*, during the focus group, PR2).

An MTA leader admitted that there was a certain kind of financial support from the municipality to realise partial interventions for the water network but, ultimately, the municipality has never agreed to recognise the community's illegal land occupation and it has always opposed their staying in that area:

> Ahia, the municipality of *Charallave*[29]! What can I say? We always had problems with the municipality, because of too much bureaucracy, too many delays. At a certain point, you are stressed because you knock on the doors four, five, six, seven, ten times and you never get an answer and you feel disrespect for the office. It is clear that they helped us, I don't want to say they don't help us, there has been effective aid, but they didn't do all that is necessary, this is what I can say … there could have been more, more communication with the community because at a certain moment the mayor came to communicate – I have a piece of press concerning it – that … there would have been an housing plan in the settlement ("*plan a vispa*") with dignified houses … an official promise by the mayor in the year 2002 … and when she promises something and she does not fulfil it, the community feels defrauded, cheated, as you promised me that, but you didn't give me that … (MTA leader terrace 11, PR2).

Direct relationships and communication between MTA leaders and local institutions were very limited in this area. Most water committee leaders had never gone to municipal or Hidrocapital offices to complain about their problems and look for solutions. Direct contact with local institution representatives had concerned just one MTA leader in *terraza* 11 (the president of the PR2 MTA). The intervention from Hidrocapital following these requests was not because of official and regular collaboration between the technical water committee and the water company in the framework of a wider and officially developed water project, but rather the result of personal, continuous, and informal pressure from the MTA leader on a specific water enterprise engineer.

Prior to concluding this section, it is necessary to remark on the municipal role in water service improvements, as there seems to be a clear gap between the theoretical collaboration among local actors foreseen by the new Venezuelan law for water and sanitation services (LOPSAPS),

---

29 *Charallave* is the capital city of the Cristobal Rojas municipality.

and the participation of local governments in the delivery of municipal services on the ground. In the view of the water company officials, the municipality has very limited responsibility. The coordinator of community participation for the Losada Ocumarito System highlighted the fundamental role of municipalities in informing the water company about low-income communities' necessities (on the basis of the complaints they had received) and in technically assisting Hidrocapital engineers, without pointing out any other particular municipal duty. The general manager of the Losada Ocumarito System focused just on the municipality's task to provide low-income communities with official settlement recognition, in order to obtain water improvements. More radically, the ex-coordinator of community participation for the Losada Ocumarito System stressed that:

> MTAs are the direct partners of Hidrocapital ... Before the mayor had the possibility to give the task to somebody else to realise a project. Now, on the contrary, a person can be handed a specific project if it is the community who decided it and if he responds to the requirements. At the beginning, mayors were scared about MTA and CCH, this thing has to be said! ... Because culturally the mayor had the power to decide, but now decisions have to be taken with the *Consejos Locales de Planificacion Publica.* The mayor is simply a presence, a person who represents people but without deciding. ... The mayor and the parish board are just managers (*gestores*) of public policies, not deciders.

Maybe surprisingly, all the interviewed municipal public servants, with the exception of the Mayor of Paz Castillo[30], fully considered that responsibility for water delivery exclusively rested on the water company. Paz Castillo municipal public servants confirmed that they did not have any contact with MTAs because the reference institution for water-related problems was Hidrocapital. The person responsible for the "Public Services" sector in the Paz Castillo municipality admitted that "if we become aware of a problem in a community, we call Hidrocapital because they are responsible for solving or repairing the problem". Finally, in the same vein, the director of "Public Infrastructure" in the Cristóbal Rojas municipality stated that "if there is a problem with the pipelines, we call Hidrocapital and they always assist the community". It is therefore not surprising that, according to these attitudes, the municipality did not develop any efficient participation mechanisms between itself and the water technical committees, and that, eventually, it did not play any significant role in water delivery.

---

30 During the interview with the Paz Castillo Mayor, confirmed that he was more aware of community problems than municipal engineers since his specific job rested on "being close to the population". The statement clearly raises the question of the feasibility of this purpose in the entire Municipality's territory.

Representative democracy

The procedure through which technical water committee leaders were elected substantially differs in our two cases, indicating that, first, no standard official practices were followed to elect MTA representatives in the settlements considered and, second, there exists yet again a positive correlation between more informal, flexible and quick procedures and concrete results in water improvements on the ground.

The selection of MTA leaders in the *Hacienda el Carmen* (where a water project was realised) has very marginally engaged the interest of community members in a direct and formal way, as water committee representatives have been appointed, through a rather informal process, based on their previous commitments to other community-based organisations or on their relationships with existing community leaders. People responsible for the first MTA – which was set up in 2003 – generally belonged to the core active groups who formed the various neighbourhood associations in the sectors. The leaders of the HeC MTA that followed were mainly people already working in the neighbourhood association, but also residents who wanted to get directly involved in finding a solution to their water problems. More open and formal community elections, according to water committee leaders, seem to have taken place in the framework of the project's development, to identify people to work out the mapping of and census in the various condominiums[31], although only one resident in 23 recalled the election of the person in charge of the condominium.

Conversely, in the settlement of *Paso Real 2000*, MTA leaders were elected through a more formal and open procedure. An assembly was organised in a central area of the settlement and the whole community was invited to participate. At this first meeting, Hidrocapital representatives were also present, in order to explain the function of the technical water committee. Membership of the MTA was open to all residents who, on a voluntary basis, were willing to participate in the water board:

> They [Hidrocapital representatives] came here and they showed us the video of the *Hacienda el Carmen* about their MTA organisation and the process looked very interesting … and starting from that moment the community decided to collaborate with us, at the beginning 36 people subscribed for the MTA, some continued but others left. We wanted to give the MTA juridical personality … the second step was the *acta constitutiva* … through a meeting amongst the people in charge of the MTA, about 26 people, and in

[31] In the *Araguaney Amarillo* sector, 4 groups of people belonging to the four main streets of this area have been elected responsible for the *Aaraguaney Amarillo* condominiums, while in the sector *Colinas* 9 people have been elected responsible for the condominiums.

> this occasion we elected within us, through the vote, the persons who had to constitute the administrative board … (MTA leader terrace 11, *PR2*).

Besides this accurate official and formal procedure, the PR2 technical water committee was no more successful in producing any significant results in water service upgrade and was instead sluggish in its operation.

6.4. Does accountability meet effectiveness in water delivery?

As with participation, the accountability of local governance arrangements involves two different perspectives reflecting the two levels of MTA involvement in improving water supplies: the responsibility relationships between the MTA and all of the community members and between the MTA and local institutions. While the latter has been mainly investigated through the calculation of the Urban Governance Index at the municipal level (especially institutional accountability to community-based organisations and citizens) and only a few additional comments need be added here, the former reports much more on the settlement level, and thus on the quality of the responsibility shown by the water committee to the entire community. The accountability of MTA leaders to community members is related, first of all, to the concept of "answerability", which means the capacity and effectiveness of MTA members to inform residents about the process of improving access to water provisions (*transparency*). Secondly, to the concept of "enforcement", meaning the MTA's capacity and effectiveness in improving water provisions for the whole community, while going beyond their particular interest in getting the service for themselves, or for a reduced area of the settlement; working with other community-based organisations for the settlement's development, outside the hierarchies and leaderships inside the community; and behaving as the spokespersons and defenders of all community needs and priorities in front of the local institutions (*integrity and responsiveness*) (see Ackerman, 2004).

Concerning transparency at the community level, when information needs to be shared among the community, this is pretty effectively communicated in a rather informal way, through paper notices or word of mouth by the few residents who participate in the public meetings organised by the water board leaders (when these gatherings exist). However, neither information about, nor involvement in the overall decisions for, the water network's upgrade seem to be organised by the MTA for dwellers, as the community's collaboration is essentially related to the technical aspects of the water project. In the case of very partial improvements, as in the PR2 settlement, this informal, irregular and selective way of managing communication within the settlement is probably one cause of the deep distrust amongst residents with respect to the technical water committee.

Looking to integrity and responsiveness, it is obvious that there is a positive connection among the presence of real and concrete improvements in the water network, MTA leaders' engagement in lobbying local institutions for the wellbeing of the whole community, and the capacity of the water committee to collaborate with other community based organisations. In the *Hacienda el Carmen*, the effective delivery of water supplies seems to rely on, beside other elements, commitment from the technical water committee representatives, but also on their participation in other community-based organisations, which allows them to operate more effectively as they receive better knowledge and information about the community. In the *Paso Real 2000* settlement, however, ineffective and inequitable drinking water supplies join the scattered actions of MTA leaders who rarely participate in other associations and often act on behalf of the small area of their terrace rather than for the whole settlement. Besides the specificities of each community, it appears obvious that improvements in water service delivery have joined up dwellers with their leaders.

With respect to accountability in the relationships between water boards and local institutions, very little information is available to MTAs from local institutions and, once more, is instead better obtained through informal and irregular communication channels. The efforts of the MTA in PR2 in formally organising the water board (public election of leaders, clear administrative board, internal regulation, etc.), and thus attempting to be more accountable towards both institutions and community, has not corresponded to better engagement between the municipality and the water company in answering the problem of water shortages.

MTA accountability to community members

Transparency

The capacity and effectiveness of the technical water committees to inform dwellers about the process of improving access to water provisions can be weighed up against community members' knowledge – and the way in which this knowledge is acquired – of the MTA and its purpose, as well as of the existence of and outputs from MTA meetings with the community. So, first, as we have already extensively shown (see § 6.3.1.), no general trends concerning residents' awareness about the technical water committees, the condominiums and their purpose can be sketched out, as there is a certain balance between those having an idea of these arrangements and those who do not. Community members who have experienced some improvement in their access to water generally tend to recognise the MTA/condominium's duties in the water provision upgrade and maintenance management. Second, information on in-progress or future water service upgrading is essentially shared between the MTA leaders and the community through assemblies. But the absence of any relevant information to be shared with the community clearly affects the existence of these gatherings and, the transparency of the MTA's actions for ensuring progress in water delivery.

When the water project progressed and assemblies were frequent, such as in the HeC settlement, community members were generally informed about the existence[32] and the outcomes of these meetings. They were usually aware of meetings about the water service and the decisions made in these assemblies through informal and occasional information provided either through word of mouth by their neighbours, or through MTA leaders (Q8a). The latter was realised via small paper notices hung up in the central and frequented sites of the settlement and by information passed through door-to-door communication (the latter system was particularly used in the areas, such as the *Colinas* sector, where the committees responsible met dwellers to collect payments for water at each household). In any case, the information's dissemination to community members was largely related to the technical aspects of the water committees. Dwellers appeared to be much more informed about and involved in the operational aspects of the water service delivery process (pipeline construction and water payments) rather than informed about and involved in the inclusive development process that the MTA should bring about in the context of a *participacion protagonica*, as highlighted by MTA leaders and Hidrocapital officials. For example, residents in this settlement generally knew the person(s) in charge of the water service (20 interviewees out of 23), but they scarcely knew the community-based organisation – MTA – to which these persons belonged (only 8 people out of 23, Q2a).

When instead there were no significant service improvements and thus no specific technical information to communicate, transparency related to the MTA was heavily affected. At the beginning of the water committee in the PR2, for example, meetings between the community and the MTA represented the most common system of communication between these two actors. But participation had always been very limited[33] and, because of the impasse in the water project, there had been

32 About two thirds of residents knew about these meetings and confirmed knowing the outputs of these assemblies (16 people out of 23 – Q8a), even though just 8 people, out of the 23 interviewed confirmed having participated in them (Q9).

33 This is witnessed by the fact that just 3 interviewed residents out of 23 remembered having participated in meetings concerning water, and by a MTA leader's comments about residents' difficulties in going to such assemblies: "It is difficult to go to the

a progressive lack of relevant information to be transferred to community dwellers from MTA leaders. The slow and uncertain water committee activity, together with a puzzling communication practice, occurring in a rather informal way has led to a serious problem concerning transparency with respect to the MTA and, in turn, to a deep distrust and suspicion toward the technical water committee and its leaders. For example, MTA leaders collected residents' signatures in order to solicit Hidrocapital and the municipality to take into account the settlement's scarcity of water. Eventually, dwellers never understood why they all had to sign given that Hidrocapital contributed to only a part of the terrace:

> There are many problems in the lower part of the terrace because, when we changed the main pipe, they bothered us and they said that we were doing the project through Hidrocapital and we didn't take them into consideration, it is not like that …the pipe couldn't reach everybody… but I think that with the project of the CTU and MTA we are going to solve the problem because we are going to urbanise. … We asked for signatures to request the project but they think that we collected signatures to solve our problem in the higher part of the settlement, but it is not like that. The problem is that in the lower part of the settlement they did not organise themselves and so they did not do the project. There was a confusion because while we were collecting the signatures, at the same time, the works started and so they though that we were doing the works with their signatures. … The signatures were for everybody, but they think that we did this work thanks to their signatures (MTA leader terrace 11, PR2).

Integrity and responsiveness

The first aspect related to the integrity and responsiveness of MTA members towards community dwellers relies on the water committee's willingness and capacity to improve water provisions for the whole community – while going beyond the particular interests of the leaders in obtaining services for themselves, or for restricted neighbourhood groups – and the leaders' commitment to accomplishing their duties honestly. It is clearly difficult to appraise to what extent community leaders engage in that process for themselves or for the whole community. But there is some empirical evidence of the fact that, if at the beginning, the

meetings when there is something that you don't like; because when I say to people that what we have been doing is with the aim to obtain the service and ensure payment, this is when people don't like it and they say 'why do we have to pay if I put the hosepipes in by myself?' So, I explained: 'yes, the hosepipes are yours, you put them in, you spent the money, but the service belongs to the water company, so, the company is going to ask you to pay for the service" (MTA leader *terraza* 11, PR2).

idea to set up an MTA is clearly attached to water service improvements for the entire community, water committee leaders tend afterwards to change behaviour. In both settlements, in effect, the technical water committees have been composed of representatives from all five sectors (in the HeC) and all 11 terraces (in the PR2), to provide water to the whole settlement. Nevertheless, once specific problems or opportunities occurred in a part of the settlement, answers were first and foremost found within the households affected rather than at the settlement level. When, for instance, the *Nueva Repubblica* sector in the HeC was experiencing a water shortage, only the committee's leaders of this sector mobilised; or, when only partial improvement was realised in a part of the *Paso Real 2000*, MTA leaders had not been able to take advantage of the water company's intervention in terrace 11 in order to lobby the company for a comprehensive project – at least in the terrace – and to provide piped water to the whole community. An exception is however represented by the payment collection difficulties present in the *Colinas* sector of the HeC and the subsequent proposal to establish a cooperative at the settlement level.

A second facet concerns the MTA leaders' collaboration with other community-based organisations in order to contribute to the whole settlement development – outside the hierarchies and leaderships inside the community. Even though the coordination among different community based organisations is currently imagined within the framework of the communal councils, at the time of the investigation, this collaboration was extremely narrow. In the *Hacienda al Carmen*, MTA leaders were often involved in other community associations, such as the *comité de tierra urbana* – CTU, the *comité de salud*, the *mesa tecnica de energia*, and neighbouring associations, but they tended to keep these CBOs separate, with the aim of not mixing their roles and tasks among the different organisations. In the settlement of *Paso Real 2000*, on the contrary, very few community leaders were involved in just one other community-based association (the urban land committee), provoking the director of the "Rural and Urban Planning" section of the Cristóbal Rojas municipality to pass very negative judgement on their capacity to get organised

> People are uninformed because they don't know what they are doing; there is no communication between the different associations; one hand does not know what the other is doing! ... they don't get on well amongst themselves to realise the projects ... this community is quite atypical insofar as people in the CTU are doing just this; in other communities people are willing to participate and put themselves in all of the associations ... therefore there is no communication between the CTU, MTA and the Communal Council in *Paso Real 2000*.

The third aspect of the water committee's integrity and responsiveness towards community members concerns their leaders' capacities to behave as spokespersons and defenders of all community needs and priorities in front of the local institutions. Also in this case, we can only rely on their accounts to appraise their willingness and commitment to personally work within the community and engage themselves in lobbying the municipality and the water company. According to MTA leaders, there is no doubt that their engagement with community members in front of the institutions (the municipality but above all the water company) is real and deep. But the empirical evidence collected on the ground nuances, to a certain extent, their proposals. In the PR2 settlement, for instance, the direct commitment of MTA leaders to local institutions has been very limited in practice, and moreover limited to the terraces where they lived rather than to the whole settlement. In the HeC settlement we have already highlighted how the water committee's leaders have involved community members in the MTA's decisions and activities only to a reduced extent (§ 5.1.2.).

What clearly appears, however, is that the leaders' commitment was a direct result of the incitement by President Hugo Chavez to assemble everyone at the community level, in order to work with institutions and obtain services. This situation has a major impact on water committees' accountability to the community: if the government-led process to improve water delivery through the MTA rests on the legal basis provided by the LOPSAPS and support from state institutions that aim to guarantee real improvements in service provisions, dwellers are also asked to conform to this procedure and organisation in order to benefit from service upgrading. In practice, we could thus argue that MTA leaders are much more requested to behave as the spokespersons and defenders of the government policy in front of community residents, rather than being more responsive to the settlement's residents.

> When we began to work, we were a *mesa de trabajo*, later there was the proposal by the president of the *mesa tecnica de agua* within Hidrocapital … So, with Hidrocapital we held the MTA meeting with the water company aid and then they called us to a national meeting … (MTA leader in the *Colinas* sector, HeC).

> I became aware of the MTA through the president; here the president taught us many things, he taught us to work with the *mesa tecnica de agua*, with the *cominté de tierra*, with the energy, with everything; he told us "organise yourself; people, you have to organise yourself in order to obtain your …", how does he say? … "I cannot do it alone, you have to help me and organise yourself in order to live better"; he was the person who taught us that, also in the facts (MTA leader in the terrace 2, PR2).

We know that water is important … I remember that we saw a programme of the president at the television (a resident in *Paso Real 2000*).

Institutions accountability to the MTA, and vice versa

Even though according to the new law on water and sanitation services, the municipalities and the hydrological companies should play a fundamental role in informing and helping communities to organise themselves into water committees, the situation on the ground differs. The information concerning how a community should organise itself into a MTA is not directly or systematically provided by Hidrocapital or the municipality. It is generally available when people go to the water company offices to complain about their water shortage (as happened in the cases here analysed) and through the state's popular programmes on the media (radio and television). The scarcity of shared information between local institutions and the MTA is also witnessed by the practically absence of regular and formal exchange opportunities with the technical water committees. In general, only a few meetings are organised between the water committee and the water company when a community, such as the HeC, is realising the water project, while meetings are totally absent with the municipalities because the service is considered the responsibility of Hidrocapital, especially in urban and peri-urban areas. As a result, important changes – such as the possibility for municipalities to directly assume responsibility for their water service, through the *mancomunidad de municipios* – are completely unknown by residents[34].

Nevertheless, the reverse accountability of the water committees to local institutions through the way in which the MTA organises itself in front of the municipality and the water company (internal regulations), and how it reports changes, problems and necessities to these institutions, is also extremely poor. For example, the HeC MTA has never provided itself with a formal organisational structure or any form of regulation – in order to project a clear image of its tasks and duties – although setting up a cooperative to organise water payments and network maintenance seems to be a first step towards a more formal and institutionalised appearance in front of local institutions and community dwellers. In

34 MTA leaders in the *Hacienda el Carmen*, for instance, were organising water payment though a forthcoming cooperative, without any knowledge of the possible changes in the Hidrocapital-municipality relationship – with respect to water service provisions – according to the new water and sanitation law. This fact is particularly relevant given the prospect, though rather unlikely, of the possible substitution of Hidrocapital with another water service provider.

Paso Real 2000, on the contrary, despite the water committee's attempts to provide itself with a formal internal regulation – written and registered at the municipality – and a clear administrative board[35], communication between the MTA and the municipality or water company were totally absent, given the deadlock situation of the project.

[35] This form of accountability seems more reliant on the desire to force institutions to provide the service, rather than on the concern to manage the water board with transparency and integrity: "they [Hidrocapital and the municipality] gave us the pipelines because we have a juridical personality and so they have to give attention to you, but if I go personally, no, nothing like that, no, only if we are organised" (MTA leader, terrace 11).

CHAPTER 7

Conclusions

This concluding chapter is divided into three parts. First, we present a comparative analysis between the institutional dimension of urban governance at the municipal level and the informal community-based governance arrangements at the settlement level, based on the common principle framework provided by the UN-HABITAT Urban Governance Index and developed through a multilevel mixed model study. The analysis shows how the new governance arrangements for water improvement in Venezuela – which are based on formal as well as informal collaborations between official institutions (represented by the municipalities and the state-owned water company) and practice-rooted community knowledge and activities represented by the water committees – can be excluded from an urban governance assessment that exclusively relies on official quantitative data and information, such as the UGI indicators. We thus contend that a qualitative analysis focusing on the extent of the communities' involvement and inclusion in governance arrangements developed around the MTA could usefully, and even should, integrate the UGI quantitative indicators to assessing the real extent of urban governance.

In the second part, findings stemming from the comparative analysis between governance at the municipal and settlement levels are grouped according to some relevant dichotomies between normative and operational challenges, decentralising and centralising tendencies, formal and informal relationships and agreements. The third part focuses on the potentialities and limitations of and possible improvements to the Urban Governance Index as a tool for measuring urban governance, not only at the city level but also at the local and settlement level.

7.1. Urban governance at the municipal and settlement levels

Comparative analysis is performed against the UGI sub-criteria for assessing urban governance (see § 4.2.) although meta-inferences have been elaborated for only six out of the eight of them. Sub-criterion 5 on "representative democracy", in fact, does not allow a direct and relevant comparison between the democratic election of the mayor and council at the municipal level and that of MTA leaders at the community level, while a few considerations have already been presented in § 5.3. and 6.3. For sub-criterion 8, "upward accountability" from the municipality towards higher levels of government is not considered here because of the incomplete decentralisation process already extensively described and, in turn, the limited room for comparison in analysing local governance for improved water provisions on the ground. A summary scheme of urban governance meta-inferences divided into the 8 UGI sub-criteria is presented in table 7.3.

Effectiveness: Subsidiarity of authority, sufficient resources and autonomy

Local governance arrangements for improving access to drinking water seem characterised by the delegation of authority at the community level – rather than at the municipal one – through the handing over of extensive autonomy and financial resources to the MTAs. Results stemming from analysis at the municipal and community levels are, at the same time, convergent and complementary in showing that responsibilities for water service operation appear entrusted to the technical water committees, rather than the municipal offices charged with it. Both levels of investigation contribute to portraying a situation where, in practice, water supply is organised and operated jointly by the MTAs and the centralised hydrological company, while the municipality – which should constitute the intermediate link between the state and the citizens – plays only a minor role (see figures 7.1. and 7.2.).

Despite a comprehensive normative framework aimed at completing and consolidating the decentralisation process initiated in 1989 and involving water and sanitation services too (LOPSAPS, 2001 and 2007), investigation at the municipal level brings to the fore a situation in which local governments have unclear powers and limited financial resources in the operation of basic services, including water. In practice, the state-owned water company Hidrocapital continues to be the only official institution supplying drinking water, working directly with low-income communities and organising the technical water committees.

This tendency is confirmed by the qualitative analysis run in the *Hacienda el Carmen* and *Paso Real 2000* settlements (see also conclusions in relation to the sub-criterion of participative democracy). Interviews with MTA leaders, municipal public servants and Hidrocapital officials confirm that, in the former settlement, responsibility for the realisation and operation of the piped network has been principally shared between the community, the MTA and the water company while, in the latter, community members and MTA leaders have mainly relied on self-made solutions to their water shortages, despite recognising the formal or informal support they receive from Hidrocapital. Despite these differences, stakeholders' perceptions of their relative importance in the water upgrade process (see table 6.2. and 6.3.) clearly show that in both cases the municipality's participation in the delivery of water provisions was deemed the lowest amongst all of the interviewed groups involved. The delegation of responsibilities from the central government to the community level is also shown by the recently instituted National Fund to finance low-income communities directly through central state allocations (see chapter 3).

The incomplete decentralisation of powers and financial resources available to municipalities, and the concurrent delegation of responsibilities and economic means to communities, makes the current local governance arrangements for improving drinking water provisions heavily dependent on central government policies. This co-production system for water services appears, in effect, strongly structured around a multitude of water committees and a public distribution enterprise. According to Carlos Mascareño[1], the dispersion of power among a great number of community-based organisations entails its concentration at the national level, represented by the public hydrological enterprises: water management is in theory decentralised to municipalities, but these are at the same time bypassed in practice by the *Mesas Técnicas de Agua*. The privileged link between low-income communities and the central power might thus reduce municipal efforts in promoting and delivering improved basic services as well as negatively affect valuable dialogue and decision making between municipal public servants and these communities (Moretto, 2010).

But the effectiveness of local governance arrangements is also affected by the specific water access conditions at the settlement level. The existence of somewhat illegal connections to the water mains in *Paso Real 2000*, although informal, unreliable and not accessible to all dwellers, together with the long-standing imprecision of the administrative zoning of the area, have represented two major limitations in the effectiveness of local arrangements, and the ensuing delivery of an effective service. This has resulted in all three actors – the community, the municipality of Cristóbal Rojas and Hidrocapital – shifting responsibility onto each other. According to the Director of "Urban and Rural Planning", for instance, the municipality could not intervene in the PR2 settlement as the area was still classified as an industrial zone and not a residential one (even if in his view the solution lies in a simple request from the community for an administrative change). Hidrocapital officials stated they could not supply water to the area until its land regularisation, and,

1 Interview with Carlos Mascareño, researcher at the CENDES – Centro de Estudios de Desarrollo of the Universidad Central de Venezuela, in 2006.

additionally, they considered that water problems were very limited in this settlement and thus not a priority. Finally, according to MTA leaders, the area had already been recognised as residential and they considered that the water company was incomprehensibly stalling for time. If any governance arrangement had previously been made, it is clear that the situation was now in a confusing stalemate.

Conversely, in the *Hacienda el Carmen*, the very poor water access conditions had unquestionably been a stronger motivator for community members to mobilise themselves and for the water company to intervene. Furthermore, the settlement's huge size and population, and hence its visibility[2], together with easier legal conditions for the public authority to deal with the land property rights, have undoubtedly contributed to making Hidrocapital actively sustain the HeC's MTA and, thus, achieving a more effective delivery of drinking water through the piped network. It must be noted however that the municipality of Paz Castillo only formally endorsed – through the presence of its mayor – the already well functioning agreements between the MTA and the water company. The municipality was not requested to mediate this relationship or practically intervene in the realisation of the piped network (apart from a limited financial contribution).

Figure 7.1. Sharing of responsibilities for water services' operation in <u>theory</u> (LOPSAPAS, 2001)

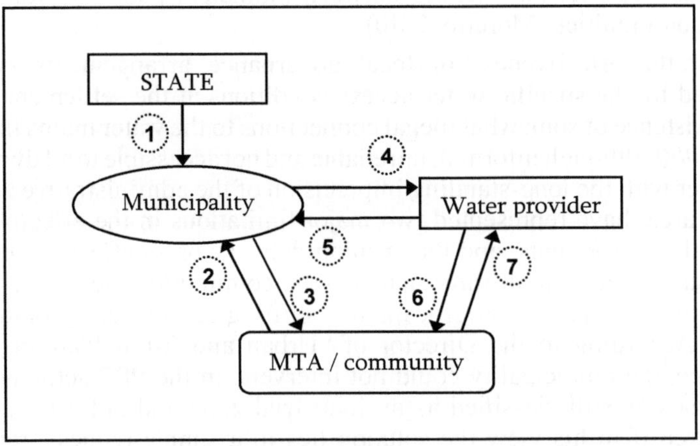

2 The HeC settlement is often used by Hidrocapital as a successful example of MTA functioning and achievements at the national level.

Figure 7.2. Sharing of responsibilities for water services' operation in practice

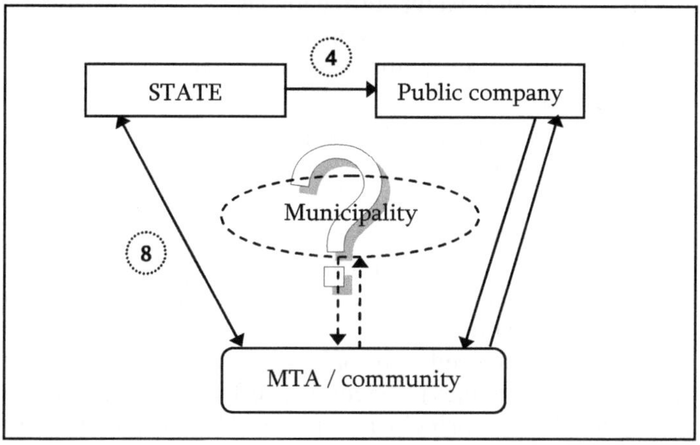

(1) Decentralisation of water services

(2) Information and communication

(3) Institutional and legal endorsement, financial contributions, facilities to receive complaints

(4) Delegation of the service operation, communication and information

(5) Provision of the service, technical and financial support, information and education

(6) Provision of the service, technical and financial support, information and education, MTA organisation and management

(7) Land and water surveys, community management (condominium), project supervision, payment collection

(8) Information, communication and education, possible financial support from the central state

⟶ Prevalent and continuous relationship

- - → Limited and *ad hoc* relationship

Effectiveness: Institutional efficiency in delivering public services and responding to civil society concerns and welfare

The effective water service provision benefited mainly from the effectiveness of governance arrangements at the community level, regardless of the overall municipal efficiency in financial management and planning as well as the institutional mechanisms that the municipality could put in place, as per the effectiveness (and also the accountability) UGI indicators. Since water and sanitation services are still operated by the water company Hidrocapital, the profiles elaborated at the municipal level do not give any evidence of the local governments' efficiency in delivering basic services. Analysis at the community level confirms that effective drinking water provision is based on collaboration between MTAs and the water company, with variable support from the municipalities – mainly through financial contributions – and community residents.

Two main considerations arise from the comparison between the municipal and community levels of analysis with respect to institutional efficiency for improved services and responsiveness to citizens' needs. Firstly, the achievement of an effective water service provision seems highly dependent on the informal relationships between the MTA leaders and Hidrocapital officials – and, to a very limited extent, municipal public servants – while formal and institutionalised interactions amongst the actors involved play only a minor role (see also point 6 on participative democracy). Secondly, effective responsiveness to water supply needs rests principally on the MTA's capacity to establish regular channels of communication with the institutions involved in the service delivery process and provide them with updated information regarding the water access conditions in the settlement, as well as on Hidrocapital's availability to work with low-income communities. In any case, responsiveness to water supply needs does not rest on the existence of municipal mechanisms aimed at facilitating communication and information (see also UGI accountability indicators).

Table 7.1. shows the discrepancy between the municipal and community levels of analysis in the two case studies. Greater capacity of local governance arrangements to effectively improve water service provisions appears, to a large extent, to be linked to regular as well as continuous communication and information flows between MTA leaders and the water company officials, often taking the form of informal contacts and relationships, rather than being connected to a higher scoring of UGI effectiveness indicators at the municipal level. In the *Hacienda el Carmen*, the whole settlement had been provided with drinking water; the MTA had been working constantly with the community, Hidrocapital and the mayor, fostering communication and sharing relevant information amongst these actors, mostly in an informal way. Nevertheless, the *HeC* settlement belongs to a municipality where the UGI effectiveness sub-index scored rather low. Conversely, in *Paso Real 2000*, despite a higher UGI effectiveness sub-index, reliable drinking water provisions were still absent. The local government and the water company were barely informed about the settlement situation, and the formal and well organised MTA operating in this settlement had produced very modest results.

Table 7.1. Effective water supply at the community and municipal level

Municipal level		Community level			
Municipality	**UGI effectiveness sub-index**	***Effective water service provisions***	**Informal contacts and relationships**	**Regular, continuous communication**	**Community**
Paz Castillo	↓ (0.31)	↑	↑	↑	*Hacienda el Carmen*
Cristóbal Rojas	↑ (0.49)	↓	↓	↓	*Paso Real 2000*

However, caution has to be used in presenting this comparison between the effectiveness of governance at the municipal and community levels, as results for the UGI calculations in municipalities have to be nuanced. On the one hand, both municipalities scored quite low in respect to the effectiveness sub-index, giving a measure of the incomplete decentralisation process and, thus, showing that urban governance mechanisms aimed at ensuring institutional efficiency in the delivery of drinking water were not in place in both cases. On the other hand, differences in the UGI effectiveness sub-indexes are mainly the result of the more efficient financial management in the Cristóbal Rojas municipality, rather than to other indicators related to service provisions. All the same, this better effectiveness in local governance (higher revenues per capita, smaller financial dependence on transfers, etc.) does not seem to correspond to a greater contribution from the municipality – financial or organisational – to the improvements of the water supply in low-income areas, such as *in Paso Real 2000.*

Equity: Distributional equity

Despite the fact that UGI equity indicators portray a situation in which drinking water provisions through piped connections are widespread in the studied areas and their accessibility is secured by pro-poor policies, local access conditions may present severe shortcomings according to the different low-income communities. Despite the relatively high percentage of households connected to a piped water network in the municipalities of Paz Castillio and Cristóbal Rojas (respectively, 74.82% and 92.21%; INE, 2001), on the ground there were very different water access conditions, not only between the two settlements, but also within the same community.

Household connections in the HeC were considered well-performing by all of the local actors as water was available one day out of two for all of the residents. Nevertheless many households received water with varied pressure and frequency according to their relative height and distance from the main pipes; while others were almost completely lacking the service because of the new illegal connections realised by neighbourhood settlements in the same mains network. In PR2, the inequity in accessing water was even worse. Some areas were provided with quite a good service, night and day; some areas accessed water through illegal connections for just 6 – 7 hours a day, in small quantities and with low pressure; other areas were totally lacking in any form of access to water and only few households benefited from Hidrocapital intervention in building pipelines. Results from the qualitative investigation at the community level complement the optimistic figures stemming from the 2001 and 2011 Venezuelan National Survey, which do not report problems related to the quality of the service and confirm, at the same time, that in low-income communities an extremely high percentage of the population does not have continuous access to water. But these results also help to disclose inequitable conditions in accessing basic services at the intra-community level, which entail spatial micro-fragmentations within the settlements (Navez-Bouchanine, 2002; Moretto, 2010) and may cause tensions and conflicts within the same communities, resulting in a situation that could prevent further development.

Anyway, improvements in drinking water provisions in the two settlements are unquestionable (especially in the HeC) and the importance of the new law for water and sanitation services for the achievement of these results is undeniable. The LOPSAPS (2001 and 2007) constitutes the normative framework for a pro-poor pricing policy as it provides the legal basis for an efficient and cost-effective water service with affordable prices – especially for low-income residents – aimed at an equitable access to water, together with the sharing of responsibilities for the water services.

In fact, the analysis at the community level shows that, firstly, in both settlements[3], construction and connection costs for low-income residents have been dramatically reduced through the networked project, as they have been shared amongst the water company, the community and – to a limited extent – the municipality. Secondly, in the only area that was facing water service payment at the time of investigation (the *Colina* sector in the *Hacienda el Carmen*), residents were charged with a minimum monthly tariff of 2700 bolivares and were waiting for the instalment of individual meters and the application of the social tariff (see § 5.2.1.). Thirdly, residents seemed largely willing to pay for getting drinking water through the piped network because – thanks to

3 We repeat that for PR2, this consideration is applicable only to *terraza* 11 as this was the only area where Hidrocapital directly contributed to the network construction.

the MTA's "information campaign" – they understood the importance of contributing to the service provision costs, and had started to see water payment as a way of being acknowledged in the settlement, to get an institutional response to their water shortage issues, and to receive assistance in tackling the network inefficiencies. Finally, interviewed MTA leaders reported being generally aware of the new law for water and sanitation services and knowing the importance of the 1999 Constitution for organised community-based associations: they considered these laws an incentive to mobilise and organise themselves as well as a guarantee that their efforts would be sustained and supported by the institutions.

It is important, however, to underline that the fieldwork was carried out in a transition period in which the whole process of drinking water delivery (from the settlement survey and water project to the payment collection and network maintenance) was not yet completed. The reaction of community members to the individual or "condominial" meters, together with their reaction to the possible tensions arising from the different attitudes toward payments, should also be assessed to confirm or refute these outcomes.

Equity: Procedural equity

Equity in the decision-making processes for improving drinking water provisions seems to be, simultaneously, the outcome of a more general political desire to support community-based development initiatives, but also of the capacities of local water committees leaders to organise themselves to obtain these improvements. The analysis at the municipal and community levels provides complementary results in understanding the conditions which favour an equitable decision making process and their relation to the effective delivery of the water supply.

In the case of water and sanitation supplies, the community involvement foreseen by the current normative framework offered by the LOPSAPS – in the context of a more inclusive participative democracy – might wear two hats. Not only ought residents be directly involved in the construction of the pipelines – through strict collaboration with the water company engineers, where local arrangements produce improved water supplies, such as in the HeC settlement – but they should also be included in the decisions made in relation to the water network project, because of their knowledge of the community's organisation and the previous mainly informal and illegal systems of access to water. In this sense, the current LOPSAPS can be regarded as an incentive for informal arrangements to provide water services (UGI indicator No. 13) within a normative framework aimed at promoting a more inclusive decision-making process and operational sharing of responsibilities.

Nevertheless, at the community level, the involvement of dwellers in the decisions concerning drinking water delivery cannot be taken for granted as this mainly depends on the water committee's willingness and capacity to include the whole community in local governance arrangements. The decision-making processes developed around the MTA seem more inclusive in *Paso Real 2000,* since community leaders working on the creation of the technical water committee involved the whole community in the assemblies organised to identify MTA members and present the water payment systems. On the contrary, in the HeC settlement, decisions concerning the preparation and realisation of the piped water network appear largely to have been made between the technical water committee and the water company, with the limited involvement of community members and limited knowledge of MTA organisation.

It is interesting to note that the involvement of community residents in the improvement of water provisions is generally characterised by an inverse relationship between effective water service provisions and the extent of their participation in the decision-making process, as well as their involvement in decisions about water payment. In particular, table 7.2. shows how, in the *Hacienda el Carmen,* the piped network was set up with residents participating in the practical realisation of the works but with very limited involvement in the decisions made in relation to the water project and without receiving extensive information about the water payment systems. Conversely, in *Paso Real 2000,* inclusive decision-making processes and the widespread dissemination of information by MTA leaders are accompanied by inequitable and ineffective service provisions. At the local level, broad information and inclusive decision-making amongst community members does not seem to be a determining factor in achieving sustainable drinking water supplies. Further considerations over stakeholders' participation in the decision-making process are also presented in relation to sub-criterion 6 on participatory democracy.

Table 7.2. Features related to equity in decision-making

Municipal (and national) level		**Community level**			
Municipality	**Incentives for informal provisions (UGI indicator 13)**	***Effective water service provisions***	**Inclusive decision-making process in the water project**	**Information about water payment**	**Community**
Paz Castillo	↑	↑	↓	↓	*Hacienda el Carmen*
Cristóbal Rojas	↑	↓	↑	↑	*Paso Real 2000*

Participation: Participative democracy

Whether formal or informal, regular or occasional, institutional or financial, municipal participation in the local governance arrangements for improving water service provisions is practically non-existent in both of the low-income communities selected for this research. Results from the two levels of analysis are convergent in showing that the municipal public servants are barely aware of the existence and activities of the MTAs in their municipality and scarcely involved in the service provisions operated by Hidrocapital – in urban areas in particular. Evidence from the case studies confirms that citizen participation in improved drinking water provisions rests almost entirely on the direct and more informal co-operation between Hidrocapital and the MTAs – a state institution and a civil society organisation.

Municipalities appear to be excluded from the local governance system for improving service delivery and from any form of involvement in MTA constitution or development. In Cristóbal Rojas and Paz Castillo, all the interviewed municipal public servants – with the exception of the mayor of the latter – confirmed that they didn't have any particular contact with the water committees in their municipalities and, in many cases, did not even know of their existence or status in their municipality. They gained much more access to information regarding water related problems through meetings with Hidrocapital representatives rather than directly through the community. No regular and formal meetings between the water committees and the municipality are organised in order to discuss water related problems. In general, when community members go to municipal offices to complain about water shortages, they are invited to contact Hidrocapital directly in order to establish an MTA and organise the water committee's activities with the hydrological company.

Investigation at the community level confirms that the joint work between Hidrocapital and the MTAs, together with the direct and mainly informal communication and information between these two actors, are far more significant for service provisions than municipal participation and contributions and – to a certain extent – even more effective than residents' mobilisation. Figures 7.3. to 7.6. summarise the extent of participation of the main stakeholders involved in the water delivery process (the state, the water company, the municipality, the MTA and the community) in the communities of *Hacienda el Carmen* and *Paso Real 2000*. The relationships between these actors are defined on the basis of the importance of their contribution to the effective provision of drinking water supplies through the piped project (principal or secondary relationships), of their formal or informal way of being carried out and of the specific kind of contribution that each actor makes (identified by the different numbers). As clearly shown by these figures, local governance arrangements for the effective provision of water supplies in the *Hacienda el Carmen* are characterised by the strong – and principally informally-based – relationships between the MTA and the water company. Municipal participation appears secondary and very limited (although the municipal contribution in ensuring accessibility to land tenure has been a determining factor for the follow up of the water project). Also in *Paso Real 2000*, the municipality has negligible importance in governance arrangements, while the bulk of relationships amongst the municipality, the water company and the MTA are generally much more formal than in the other settlement. In general, drinking water supplies continue to be provided in an illegal way by community residents themselves, without any relevant collaboration between the water company and the municipality and without relying on MTA leaders, demotivated by the lack of outcome of their efforts.

Local governance arrangements developed for improving drinking water supplies, however, do not just exclude the municipality, but the whole community too, at least to a certain extent. Although the community members' role is acknowledged as very important by all the stakeholders (see tables 6.2. and 6.3.), the fieldwork revealed that community involvement appears quite limited in governance practices. For example, the contribution of residents is judged more important in PR2, where dwellers solve their water shortage by themselves, rather than in the HeC settlement, where the achievement and management of the water project was mainly carried out by the MTA in collaboration with Hidrocapital. Whether providing the service by themselves or in collaboration with the other actors in practice, residents appear in any case excluded from the decision-making process.

Figure 7.4. Stakeholders' participation in *Paso Real 2000*: principal relationships

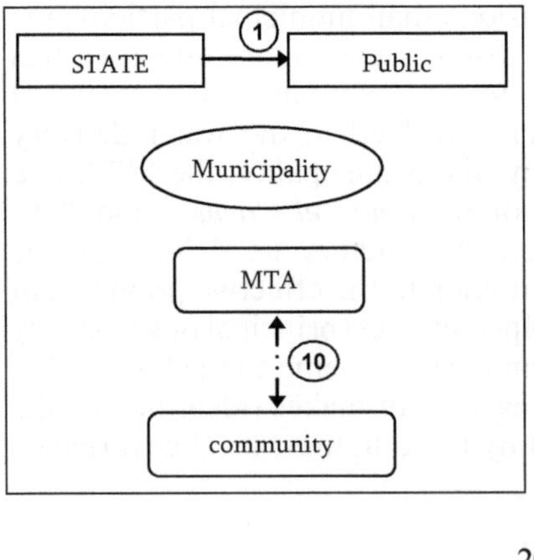

Figure 7.6. Stakeholders' participation in *Paso Real 2000*: secondary relationships

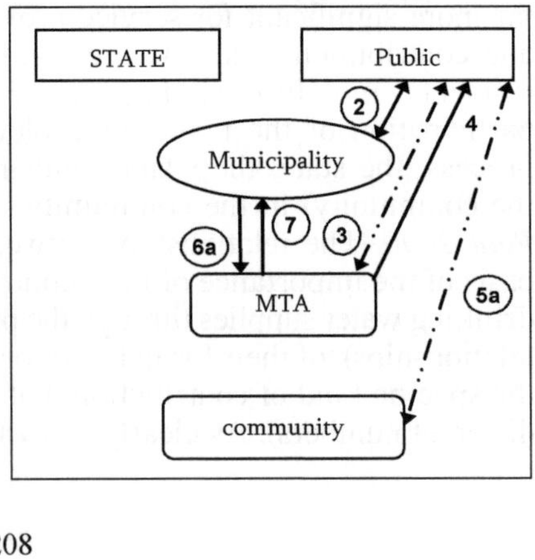

Figure 7.3. Stakeholders' participation in *Hacienda el Carmen*: principal relationships

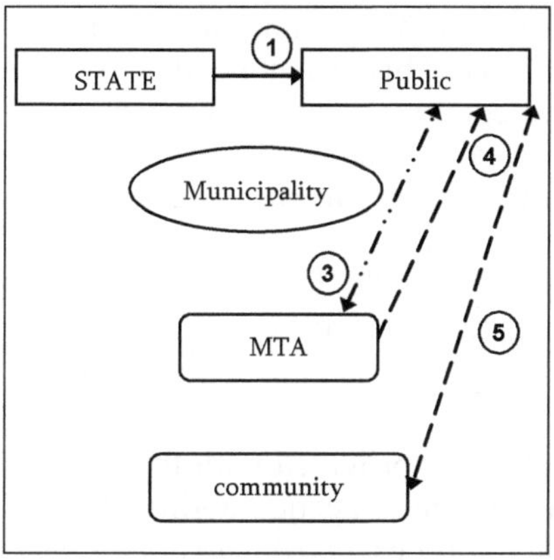

Figure 7.5. Stakeholders' participation in *Hacienda el Carmen*: secondary relationships

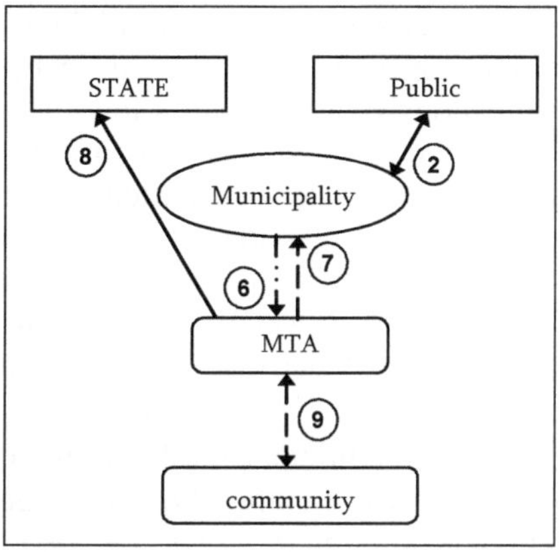

Caption:

- (1) Delegation of water service provisions
- (2) Weekly or monthly meetings between Hidrocapital and municipalities in the Tuy Valley
- (3) .Joint work between the MTA and the water company to realise the water piped network
- (4) Requests from MTA leaders (not all of them) for Hidrocapital interventions + information
- (5) Joint construction of the piped network in the whole settlement
- (5a) Joint construction of the piped network in just one terrace out of 11
- (6) Partial financial contribution to the pipelines construction, institutional endorsement of the water project, and land property rights recognition
- (6a) Very limited financial contribution and no institutional support
- (7) Requests from MTA leaders (not all of them) for municipal interventions
- (8) MTA participation to the regional and national water meetings
- (9) Technical meetings for organising the network construction
- (10) Initial meetings for the setting up of the water project (stalemate due to the lack of topographical survey)

——▶ Formal relationship

– –▶ Informal relationship

·—·▶ Both formal and informal relationships

Accountability: Downward Accountability

Accountability in local governance arrangements to improve drinking water provisions can be undermined by sluggish organisation and management at the municipal level and by the unclear division of responsibilities among local institutions). This often results in their unresponsiveness towards low-income communities and the distrust of the latter towards local institutions. Again, despite a broad-spectrum normative framework aimed at ensuring transparency, fiscal control, integrity and responsiveness and the new law for water and sanitation services, municipalities are not yet responsible for providing water services, while MTAs and community members have a difficult time getting their voices heard by local governments and, sometimes, by the state-owned water company as well.

At the municipal level, Cristóbal Rojas is provided with the specific facilities to ensure the local government's accountability toward citizens (*Oficina de Participación Ciudadana y Contraloría Social de la Contraloría Municipal del Municipio Cristóbal Rojas*), while Paz Castillo is not. However, as water services are still centralised at the national level through the hydrological companies, users' charges are neither set nor controlled by the municipality and citizens' complaints concerning water supplies are not processed by municipal public servants who just refer communities to the water company. This situation clearly raises the issue of who should be accountable to citizens for the delivery of equitable and efficient water services, given the inconsistency between responsibilities at the normative and legal level – as per the new LOPSAPS – and the practical level.

Analysis at the community level seems to confirm that the incomplete and unclear delegation of responsibilities for water and sanitation services brings about uncertainty among MTA leaders and community members, with respect to who should be responsible for improving drinking water provisions, and some confusion concerning the MTA's role. In particular, in *Paso Real 2000* – where water supplies are largely ineffective and inequitable – the MTA has been lobbying the municipality (above all for the tenure of the area) and the water company (especially for the topographical survey) without getting a clear answer from anyone concerning their water shortage and without knowing how to proceed in order to get things moving again. As previously noted, the stalemate over drinking water provisions that this settlement experienced was not going to be easily solved, as all of the stakeholders were waiting for the others to make their move.

Table 7.3. Summary scheme of urban governance meta-inferences

Criteria	Sub-criteria	Level 1 (Municipality, qualitising, triangulation)	Level 2 (Community, multi-methods, frequency effect size)	Meta-inferences
EFFECTIVENESS	**1. Subsidiarity of authority and sufficient resources and autonomy (decentralisation process)**	1M. Incomplete decentralisation process: comprehensive normative framework but incomplete administrative and fiscal procedures	1C. Scoring exercise at the community level as a proxy of the decentralisation process, showing that MTA and Hidrocapital are the two main actors while the municipality is almost absent.	1. Subsidiarity of authority: autonomy at the community level rather than at the municipal one; water services' responsibilities seem much more decentralised at the community rather than at the municipal level (→ **convergence of results; complementary results**).
	2. Institutional efficiency in delivering public services and responding to civil society concerns and welfare	2aM. Effectiveness sub-index, measuring "formal" effectiveness is higher in PR2, although it is higher for the indicators NOT related to water services.	2aC. The more effective water provisions directly relate to (A) informal contacts and collaboration between MTA leaders and mainly Hidrocapital and (B) responsibilities delegated to the MTA.	2a. Greater effectiveness of informal rather than formal arrangements supported by the non-correlation between effectiveness levels at the 2 different units of analysis in the 2 case studies (→ **divergence of results**)
		2bM. Lack of service delivery, and thus responsiveness, from the municipality as a result of the incomplete decentralisation process.	2bC. Responsiveness from MTA to community resting on effective communication and information sharing with local institutions.	2b. Responsiveness to civil society is carried out much more by the MTA and Hidrocapital rather than the municipality (→ **complementary results**)

Criteria	Sub-criteria	Level 1 (Municipality, qualitising, triangulation)	Level 2 (Community, multi-methods, frequency effect size)	Meta-inferences
EQUITY	**3. Distributional equity**	3aM. High percentage of household connections through a piped networks in both areas.	3aC. Very different access conditions between the two settlements but also within the same settlements.	3a. Qualitative analysis confirms that data has to be collected at a lower scale. Importance of equity at the community (micro) level because it could entail tensions within the settlement (→ **divergence of results; complementary results**)
		3bM. The LOPSAPS provides the legal basis for an efficient and cost-effective water services, affordable prices, new culture of rights and responsibilities (on the basis of mutual agreements between Hidrocapital, the MTA and the municipality)	3bC. Despite some differences between the two settlements, costs are shared between the municipality, Hidrocapital and the community; prices are affordable, where paid for; there exists a willingness to pay. The law is a motivation and incentive for MTA	3b. Existence of a pro-poor pricing policy and results on the ground (→ **convergence of results**)
	4. Procedural equity	4M. There is an explicit effort to better organise and further develop informal connections without removing them … LOPSAPS as a normative framework is aimed at regularising and institutionalising informal practices	4C. Decision-making processes developed around the MTA are more inclusive in the PR2 settlement (the whole community has been involved in the assemblies). In the HeC settlement, decisions are largely taken between the MTA and Hidrocapital, with limited involvement of community members.	4. Procedural equity based on the support for informal procedures at the national level and MTA capacity of involving residents at the community level (→ **complementary results**)

Criteria	Sub-criteria	Level 1 (Municipality, qualitising, triangulation)	Level 2 (Community, multi-methods, frequency effect size)	Meta-inferences
PARTICIPATION	**5. Representative democracy**	5M. Low voter turnout at the municipal level and high voter turnout at the national level	5C. Differences between the two settlements: HeC more informal and unclear (condominium and MTA); PR2 more formal and structured.	5. (→ **limitations in comparing results**)
	6. Participative democracy	6aM. Municipality as the excluded part, while the main actors are the MTA and Hidrocapital; municipality not informed, no regular and formal meetings	6aC. HeC: participation of MTA, Hidrocapital, Municipality in informal way; PR2: limited relationships and more formal, larger contribution from community members	6a. The participation of the municipality is very reduced, relationships are largely informal (→ **convergence of results**)
		6bM. No information at the municipal level	6bC. HeC: residents participate in practical actions, poor info; PR2: residents know the MTA and are quite well informed about the situation.	6b. no information at the municipal level
ACCOUNTABILITY	**7. downward accountability**	7aM. Concerning water: who is accountable? (Hidrocapital or the municipality? users' charges 20B and facilities to complain 22)	7aC. MTA members are not so well informed by institutions; no regular meetings	7a. Who is accountable to citizens for water supply? In practice it is clear, in theory it should be different (→ **complementary results**)
		7bM. There exists a normative framework: municipal bodies and CLPP for accountability; municipal willingness and capacity to set up these bodies and involve citizens	7bC. Informal and irregular communication and information from the MTA to local institutions	7b. Accountability not well structured and supported at the operational level (→ **complementary results; convergence of results**)
	8. upward accountability and accountability at the community level	8M. Not considered because of the incomplete decentralisation process for water services	8C. Only the MTA in PR2 elaborated an internal written regulation and organised itself through an administrative board	8. not considered at the municipal level

Reduced accountability from institutions towards low-income communities is also exemplified by their poor communication and information sharing. In both settlements residents were not informed by local institutions (the municipality or the water company) about the possibility of constituting a water committee: they only found out by going to Hidrocapital offices or through the state's popular programmes aired in the media. No specific and codified rules were followed in the MTA's establishment and its subsequent activities, as no regular and formal meetings were organised amongst the key stakeholders. Local accountability seems to be strongly dependent on the willingness (and interest) of the municipality and water company to respond to citizens and their complaints.

7.2. Placing urban governance for water services in perspective

Normative and operational challenges

Recent developments in the water sector in Venezuela have brought about relevant changes and improvements in the provision of the service. From a normative and institutional point of view, the new LOPSAPS provides the legislative framework, first, for managing water and sanitation services through shared responsibilities amongst the actors involved in the service delivery process. Second, for providing drinking water through a pro-poor pricing policy that entails reduced construction and connection costs and a social tariff. Third, for establishing and regulating citizens' participation in accessing water supply through the technical water committees. From a practical and operational perspective, local governance arrangements built around the MTAs have been proven to contribute to improving drinking water supplies for low-income communities and in mobilising these communities in order to develop better access to basic services and to the decision-making process (see also Cariola and Lacabana, 2005a, 2005b, 2008; Allen, 2012, 2014; McMillan *et al.*, 2014).

The evidence from the investigation, however, shows that there is a discrepancy between formal prescriptions and their implementation; between the normative apparatus and the existing *modus operandi.* At an official and legal level, several measures and mechanisms aimed at ensuring good governance for the provision of water supplies exist. They rest on the mechanisms, procedures and institutional bodies proposed within the decentralisation process, in the search for effectiveness and accountability at the municipal level; on the measures taken for less expensive and more reliable conditions of access to water supplies for low-income communities, which should enhance equity in service access; on the institutionalisation of citizens' involvement in the practical as well as decisional processes linked to the service provisions, aimed at improved equity and participation in governance. All this has been promoted through the numerous laws that the Chavez government has approved since 1998 (beginning with the 1999 Constitution) and the variety of community organisations that were created in the same period (such as the technical water committees, the community councils, the communes, etc.).

Nevertheless, at the operational level, these measures are often seen as being accompanied by their sluggish and incomplete implementation on the ground. The decentralisation of water services is not in place and community residents are often given major duties and responsibilities with scarce prospect of getting support from local governments. Facilities to make local institutions accountable to citizens as well as procedures to make MTA leaders accountable to community members are often not yet established or defined. The quality and extent of residents' participation in improving water supplies, as well as equitable access to piped water for all of the households in the settlement highly depends on the capacity and willingness of MTA leaders to really involve community members in this process and, thus, it cannot be taken for granted. Citizen participation relies to a great extent on the MTA leaders' collaboration with the water company, rather than on a fully comprehensive involvement of the whole community into the decision-making.

For local governance arrangements built around the MTA to be effective in the delivery of efficient and equitable drinking water provisions, real community involvement and genuine participation in the service delivery processes, together with the local institutions' willingness and capacity to intervene and collaborate in this process, are both essential requirements. Unfortunately, this research seems to confirm that priority has been given to the mobilisation of people, as some authors such as Wagner (2004c) and Brewer-Carías (2010) suggest, rather than to the operationalization of the comprehensive normative apparatus aimed at ensuring local governance. This gap between codified norms and local actions, together with the central government's political influence on, or rather control over, citizen participation (Penfold, 2010; Cordoba Jaimez, 2012) can easily undermine the viability and long-term run of these locally-based experiences for service provision and the wider national "missions".

Decentralisation and centralisation tendencies

Despite existing norms foreseeing and regulating the decentralisation of water and sanitation services from the national to local governments, the municipality is the missing link between the central state and citizens in local governance arrangements. Responsibilities for service organisation and management are delegated at the community level rather than at the municipal one; responsiveness to community members relies on MTA leaders' and Hidrocapital's efforts to improve service delivery rather than on municipal involvement; the participation of municipalities in governance arrangements is almost non-existent, especially in terms of institutional support and information sharing. The consequences of the recent forms of institutionalised co-production between the state water company, Hidrocapital, and organised groups of citizens in the water committees have proven to effectively overcome local government delays and/or the obstacles to setting up organisational changes and, to a certain extent, to producing a faster and less complex water supply delivery, which responds well to the diversity of local needs and operational situations. However, the excessive simplification of and the reduction of the actors involved in this process may lead to an unhappy partially inclusive approach, where the closest official elected institutional body to the citizens (the municipality) is excluded from the governance process and finds itself in the uncomfortable position of being unaccountable to dwellers. The dilution of public accountability through the co-production of services (Bovaird, 2007) can, once again, undermines the viability of these experiences in the long run. In the case on decentralised services to local governments, it seems appropriate to question whether institutionalised co-production – as described by Joshi and Moore (2004) – can really be a sustainable alternative for basic service delivery.

Besides the effects of the incomplete delegation of responsibilities to municipalities for water services, the tensions between decentralisation and centralisation processes in the water sector draw a clear picture of the political situation at the national level. As pointed out by many authors (Levine, 2002; Borgucci, 2003, 2013; McCoy and Meyers, 2004; Gil Ypes, 2004; Mascareño, 2006; Moreno, 2008; D'Elia, 2010; Peraza, 2012; Ríos *et al.*, 2012), relying directly on the mobilisation of organised groups of citizens in order to access basic services is largely aimed at strengthening the link between citizens and the central power, mainly taking the form of President Hugo Chavez. Joshi and Moore (2004: 44) have already highlighted how institutionalised co-production can be politically attractive as "states can provide services at a reduced costs and resources, and gain political support in exchange for some loss of control and power". In the case of Venezuela, it seems reasonable to wonder whether the current institutionalised co-production of water services could be considered a genuine alternative to dominant governance modes (hierarchies or markets) or if it still belongs to a state-centric approach (as suggested by some authors such as Cordoba Jaimez, 2012; Peraza, 2012; Borgucci, 2013), where President Chavez has ensured himself and his government political support by entrusting small pockets of power to a multitude of citizen organisations. Evidence from this analysis shows – at least to a certain extent – that recent governing practices to improve access to water supplies appear strongly dependent on the central state politics.

Formal and informal relationships and agreements

In order to develop co-production between the state-owned water company and the water committees, informal relationships, engagements, agreements and means of information and communication are generally preferred to more bureaucratic ones, as they appear to be far more effective than the more formal and official instruments. Drawing on the evidence of the two case studies, where drinking water has been effectively provided (such as in the *Hacienda el Carmen*, in the municipality of Paz Castillo), co-production largely benefited from informal, tentative, irregular and unofficial relationships between the stakeholders involved in the service delivery process. On the contrary, where the service was still lacking, relationships amongst the key stakeholders appeared stuck and slowed down by a more formal and loaded approach. Formal governance mechanisms appear to be less influential than the way governance processes work in practice in a rather case-specific and locally-influenced context, probably because informal and mutual agreements require less capacity than more formal procedures (Batley and Mcloughlin, 2010).

Research on service provisions in the Global South continuously acknowledges the important role of informal water and sanitation provisions and is increasingly focused on the ways in which informal and community-based practices can articulate to formal systems to provide improved services, through new and alternative forms of governance among the variety of service delivery actors[4]. The example of the Venezuelan MTAs adds further insights to this debate, especially with respect to state-community service provisions. So far, in effect, collaboration between low-income residents and the state company Hidrocapital has been reliant, first, on the informal and unofficial solutions produced by communities to improve drinking water delivery. Second, on informal agreements between these actors, developed through service co-production mechanisms and resulting in the local governance arrangements we have here extensively described.

4 Look to these authors for relevant pieces of research: Allen *et al.*, 2006; Phumpiu and Gustafsson, 2008; Olivier de Sardan, 2009; Batley and Mcloughlin, 2010; Booth, 2011a; Gaventa and Barrett, 2010; Wild *et al.*, 2012.

Nevertheless, although their articulation to a more formal and institutionalised service co-production model has been determined by the recent law on water and sanitation services, they seem, in practice and to a certain extent, to be failing to have widespread effective and equitable effects. The analysis has shown that such informal/formal relationships provided the potential for better service provisions but also carried the possibility of deadlocks in service improvements, mutual distrust between citizens and the water company (also within the same community), and the bypass of the municipality – the closest official institution to citizens.

7.3. Applying the UN-HABITAT Urban Governance Index

Experience with the assessment tool

The UGI's application to the urban governance arrangements developed in Venezuela for improving water supplies has been carried out through several adaptations, through which it was possible to adjust and further rework the Index with respect to the specific case study's context. The first revision concerned the Index's level of application and its disaggregation. The UGI was adapted to two specific municipalities – rather than to the whole metropolitan area – according to the idea that "it is at the local level that the quality of governance affects the citizen the most. Local governance indicators can thus play a very important role in assisting municipalities to improve their function in specific areas, engage more closely with communities, and become more responsive and accountable" (Narang, 2005: 4). Drawing on the calculation of the UGI indicators in Paz Castillo and Cristóbal Rojas (chapter 5), results show that these two municipalities present quite distinct degrees with respect to institutional efficiency in the local government's financial management and planning and to the mechanisms in place for ensuring accountability towards citizens. This outcome thus confirms the relevance of the UN-HABITAT's claim for imperatively disaggregating the elaboration and calculation of governance indicators between national, sub-national and local levels.

The most important adaptation concerned an experimental review of the UGI principles and indicators in order to elaborate a set of queries for assessing how the urban governance arrangements aimed at improving access to the water supply for the urban poor on the ground, at the community level (chapter 6). This is in line with a move from a top-down to a bottom-up approach in governance evaluation, but at a very local level where "citizens assess how well [their] needs and concerns are being met" (Wilde, 2011: 41) not only with respect to the criteria stated elsewhere, but according to their own principles and priorities[5]. By translating the Urban Governance Index's principle framework into a more qualitative and settlement-based investigation, the UGI has proved capable of providing a clear and comprehensive structure to shape analysis at the community level, in two main regards. First, this operation helped to structure the data collection for exploring the more informal side and community aspects of local governance systems and, hence, of the degree of involvement and inclusion of low-income communities in these governance arrangements. Second – and most importantly – it allowed for mixing the outcomes of the two units of analysis, on the basis of a common theoretical and methodological framework.

Finally, the UGI has also been reworked according to the specific governance policies, mechanisms and practices for drinking water supplies. This adaptation has included both the municipal and community units of analysis, allowing the comparison of results from these two different units. At the municipal level, the focus on water provisions has required the adaptation of those indicators concerning basic services to water supply. This allowed us to appreciate the extent of decentralisation in the water sector and the principal relationships amongst the key actors involved in the service delivery process. At the community level, the qualitative investigation concentrated on the assessment of the local governance arrangements aimed at improving water supplies within two specific settlements. Given the importance that the UGI attributes to a "principles framework" that emphasises "the quality of relationships and processes between key stakeholders at the local level" (UN-HABITAT, 2004a: 18) – rather than a framework based just on the quantity and performance of local government service delivery – it has been possible to investigate and compare governance mechanisms built around the water delivery process with respect to their dynamics and processes at both the municipal and community levels.

The Index has thus proved to be a flexible, handy and helpful instrument, adaptable and applicable to diverse contexts, levels and approaches. But despite the abovementioned adaptations, the UGI in

5 Wilde (2011) considers the Urban Governance Index as a "third generation" assessment tool, defining this category as explicitly addressing a bottom up process in governance assessments, which is supposed to be inclusive, transparent, and more democratic. But our analysis of the UGI has not focused on the process of governance evaluation (data collection was in effect not participatory as UGI calculation was performed in the context of a research study) but rather on the structure and principle framework of the evaluation tool. It is in this sense that we think that an explicit focus on the settlement/community level of governance could further contribute to make governance assessments tools, such as the UGI, really more inclusive, transparent, democratic and participatory.

itself, in the way it is presented by UN-HABITAT, cannot be considered an accurate tool to investigate inclusiveness in governance mechanisms, as the very most local level of governance, which is represented by the settlement and its inhabiting community, is excluded from UGI indicators. This is particularly striking when we know to what extent access to basic services and inclusion in local decision-making can vary, not only among settlements within the same municipality, but also within the same settlement (see § 6.2. and 6.3.). The following and final paragraph presents the main critical points that this study encountered in verifying the UGI's applicability to the case studies, and the resulting inputs for future research.

Policy implications and input for future developments

The UGI's focus is on quantitative data collected at the municipal level. As constantly observed in this book, this choice condenses governance assessments to the factual information available through city administrations. Nevertheless, comparing outcomes from the two different units of analysis disclosed that results gathered from the analysis at the municipal and community levels have been divergent, convergent or complementary. In some cases, outcomes related to the four core UGI principles and stemming from the study at the community level have been consistent with those arising from the municipal level of analysis. In some other cases they have portrayed different aspects, mechanisms and processes of the same governance sub-criterion. Regardless of the specific and changing correlation between the inferences at the municipal and community levels, it is unavoidable that the Urban Governance Index appears to fail in comprehensively appreciating the very local dynamics of urban governance for improved services. Three main considerations arise for the current and future application of the UGI.

Settlement unit of analysis

Complementing the UGI with an assessment at a more local and community-centred level constitutes not only a possibility foreseen by the Index (UN-HABITAT, 2004a) but also a necessity. Evidence from this research reveals the possible gap between improved service provisions and enhanced urban governance. On one hand, most improvements in accessing water supplies for the urban poor are directly linked to the mobilisation of communities and their commitment to applying pressure on local institutions, rather than to municipal initiatives. In assessing the quality of urban governance processes and relationships, the extent of the whole community's inclusion in the local agreements and the quality of the relationships among local actors on the ground – including community residents – cannot be excluded.

On the other hand, improvements in the delivery of drinking water can be linked to different dynamics than those directly related to the local administration performance and decentralisation levels in service delivery. On one side, better water access for low-income communities does not always depend on a strong representative democracy at the municipal level, on the establishment of accountability mechanisms within the local administration, or on the presence of a pro-poor pricing policy. On the other, public institutions (e.g. the central government or public water enterprises as in the Venezuelan water sector) may substantially contribute to improved water provisions. The UGI's focus on local governments, as the unit ensuring effective local governance, can result in a limited understanding of why some small-scale, informal and community-based experiences work. UGI assessment is thus likely to exclude those local governance practices which achieve relevant improvements in basic service provisions, involve actors belonging to the public, private (in its diverse variants) and community sectors but, at the same time, place less emphasis on the municipal level – as in the case of water services in Venezuela. Further development of the UGI should more systematically promote the integration of urban governance assessments at the local level of analysis, in order to thoroughly investigate how governance works in practice: primarily, at a community level and, subsequently, in relation to institutional levels other than local governments.

Qualitative methods

A qualitative approach to exploring local governance arrangements at the community level appears to be very constructive when looking into the more informal and practice-based relationships among the local actors involved in the water delivery process, for two main reasons. On one hand, it actually reveals aspects of the local governance arrangements that the simple UGI indicators might not disclose, extracting more information from the quantitative data[6], while at the same time double-checking findings from the calculation of the Urban Governance Indicators. On the other hand, a qualitative approach provides an appropriate framework to explore informal and mutual governance agreements that are in reality often excluded from a more quantitative methodological approach that looks for formal and easily quantifiable information: relying on a qualitative analysis can often be the only method to explore the extent of community involvement and inclusion in the governance arrangements.

[6] For instance, this has been the case with the analysis of the decentralisation process in Venezuela to understand the reasons and significance of most of the UGI effectiveness indicators.

The narrative profiles previously presented – based on qualitative research at the community level – actually fit well into some of the shifts identified by Hydén (2011: 17) in governance assessments over the last two decades, notably "from numerical indicators to narrative trajectories", as the objects of governance evaluations are always site-specific, and from an "exclusive reliance on quantitative data ... [to] a greater readiness to adopt and use qualitative methods of analysis".

Nevertheless, integration of qualitative approaches into the UGI could face severe limitations. Primarily, qualitative surveys are generally built according to unique contexts and needs and, hence, hardly transferable to other circumstances. It would be difficult for them to meet the "universality" criterion established by UN-HABITAT, but also in some cases the criteria of "relevance" and "ease of collection". Next, comparisons between quantitative and qualitative data should be run on one of these two approaches. Quantifying qualitative information and aggregating it into more synthetic indexes or sub-indexes might be in contrast with the very nature and scope of a qualitative approach, which is to provide the most detailed portrayal as possible. Qualitising on the contrary quantitative information (Tashakori and Teddlie, 2003, 2010) poses the risk of blurring the UGI objective to "enable cities to objectively measure the quality of local governance" (Narang, 2005: 1).

Normative framework

The results stemming from comparisons between the municipal and settlement units of analysis raise a more general and conceptual question about the significance and the adequacy of the UGI principle framework to guarantee inclusiveness and citizenship, as put forward by UN-HABITAT, in both service provision and decision-making. Although "all [governance] assessments have a normative bias otherwise it is not possible to judge what is good or bad or if something is improving or deteriorating" (Wilde, 2011), the analysis here presented has shown how citizens' access to better basic services and decision-making has often been reached without and/or outside governance systems that fulfilled the principles of effectiveness, equity, participation and accountability. Two major outcomes from research confirm this assumption.

First, a gap exists between the material and effective improvements in water delivery and the high score of urban governance indicators. Progresses in service provision were much more linked to the local capacity (especially from the water community-based associations and the public water company) to elaborate rapid, operational and appropriate solutions to water shortage – often in informal and unorthodox ways – rather than to high UGI sub-indexes' scores. Incomplete service decentralisation, and the resulting very low scores on effectiveness indicators, or the absence of specific municipal bodies for local administration transparency, responsiveness and integrity to ensure downward accountability in urban governance, did not prevent locally-developed and community-based governance arrangements from making structural improvement in water access. The other way round, good equity indicator scores (related for instance to the existence of a pro-poor pricing policy, of cheaper water in poor settlements or incentives for informal agreements to improve water) did not stop community residents from experiencing deep differences in the quality, frequency and reliability of their water access, entailing socio-spatial micro-fragmentations within the settlement.

Second, a gap also exists between water delivery upgrading and the extent to which local governance arrangements are inclusive. For example, effective water improvements were actually reached where community member participation in the decision-making process – and thus in local democracy – was limited, where community members took part in the realisation of the network but were generally excluded from time-consuming meetings, or again where water committee leaders engaged directly with local government representatives to get a quick response to water shortage issues rather than spending time undertaking official procedures to assure formal upward accountability. These grounded cases raise some more general questions about the capacity of the principles of effectiveness, equity, participation and accountability to really apprehend the way in which governance works in practice. Let's discuss two examples.

Looking for instance to the participatory dimension of governance, it seems appropriate to wonder whether it is legitimate to link it to democracy. We know that the number of civic associations cannot be a measure of participative democracy when the requirement for citizens to be grouped in associations that can engage the state could risk excluding at the very local level the less well networked and organised groups, the powerless voices within social groups and also the individuals outside these groups (see for instance Maloutas and Malouta, 2004; Batley and Mcloughlin, 2010; Verschure *et al.*, 2012; Lombard, 2013). We also know that in certain political contexts, such as in Venezuela, the autonomy and independence of certain CBOs cannot be taken for granted (see for instance Lopez Maya and Lander, 2011) as "participation does not necessarily lead to anything we would recognise as democratisation" (Smilde, 2011: 17), and service co-production runs the risk of not really challenging conventional forms of service delivery through grassroots organisations (as suggested by Mitlin, 2008). Participation is certainly broader than what is proposed in the UGI indicators. As recommended by some scholars (Cabannes, 2010; Hickey, 2010; Wilde, 2011), the participatory and bottom-up dimension of governance should probably include a "right-based approach", translated not only into the right to participate in governance arrangements, but also to evaluate them[7].

Turning to the example of accountability, evidence exists from research that stresses how the boundaries between the spheres of the local stakeholders taking part in the "tiers of governance" (for public services) are blurred and imprecise (Bovaird, 2007; Benequista and Gaventa, 2011; see also Oia, 2011). While institutionalised co-production of urban services appears to be a promising alternative to urban water supply, research results show that these mutual agreements between the water company and the water committees mainly developed outside formal procedures, codified rules and official administrative structures and bodies, mixing the roles of these two actors in the service delivery process. The question thus arises of how institutionalised co-production could ensure credibility and accountability through "undefined, informal and renegotiated almost continuously" relationships between the state and citizens (Joshi and Moore, 2004), rather than through clear and enforceable contracts (see for instance Ostrom, 1996; Batley, 2006; McDonald and Ruiters, 2012). Put another way, can formal, explicit and direct accountability as defined by UN-HABITAT encompass the increasingly undefined, informal and blurred relationships among local actors in service provision?

The divergence of reality from normative prescriptions when governance operates on the ground was recognised by Joshi and Moore (2004). It seems, fortunately, that "today governance relies less on normative blueprints and more on practical experimentation" (Hydén, 2011: 19). Governance assessments – including the Urban Governance Index – should as a consequence critically question the relevance of a normative and ideological framework in understanding how governance woks in practice, and systematically integrate a more qualitative, community-level and context-specific dimension into governance evaluation, to apprehend the very local dimension of governance on the ground.

7 Referring to a case in San Salvador during an Expert Group Meeting on Localising the Habitat Agenda Indicators: New paradigms in developing indicators (UN-HABITAT and Al-Madinah Al-Munawarah, 2010: 8), Yves Cabannes focused for instance on the "right to participate", while underlining that this includes "the right to request and receive information, the right to consult and propose, the right to participate in the decision making process, the right to co-management, the oversight right and the right to denounce".

Bibliography

Ackerman, John, 2004, "Co-Governance for Accountability: Beyond 'Exit' and 'Voice'", *World Development*, Vol. 32, No. 3, pp. 447-463.

Albrechts, Louis, 2013, "Reframing strategic spatial planning by using a coproduction perspective", *Planning Theory*, Vol. 12, No. 1, pp. 46-63.

Allen, Adriana and Adrian Atkinson, 1998, *The Urban Environment in Development Cooperation: A Background Study*, Office for Official Publications of the European Communities, Luxembourg.

Allen, Adriana, Julio Davila and Pascale Hofman, 2004, *Governance and access to water and sanitation in the metropolitan fringe: an overview of five case studies*, N-Aerus Annual Conference, 15-16 September 2004, Barcelona, Spain.

Allen, Adriana, Julio Davila and Pascale Hofman, 2006a, "The peri-urban water poor: citizens or consumers?", *Environment and Urbanization*, Vol. 18, No. 2, pp. 333-351.

Allen, Adriana, Julio Davila and Pascale Hofman, 2006b, *Governance of Water and Sanitation Services for the Peri-urban Poor*, A Framework for Understanding and Action in Metropolitan Regions, Development Planning Unit, UCL, London.

Allen, Adriana, 2010, "Neither Rural nor Urban: Service Delivery Options That Work for the Peri-urban Poor", in Kurian, M. and P. McCarney (eds.), *Peri-urban Water and Sanitation Services. Policy, Planning and Methods*, The Netherlands, Springer, pp. 27-61.

Allen, Adriana, 2012, "Water provision for and by the peri-urban poor: Public-community partnerships or citizens co-production?", in Vojnovic, I. (ed.), *Sustainability: A Global Urban Context*, East Lansing MI, Michigan State University Press, pp. 309-340.

Amis, Philip, Richard Batley, Jo Beall, Nick Devas, Ursula Grant, Nazneen Kanji, Diana Mitlin, Fiona Nunan, Carole Rakodi, David Satterthwaite, Cecilia Tacoli and Elisabeth Vidler, 2001, *Urban Governance and Poverty: Lessons from a Study of Ten Cities in the South*, The School of the Public Policy, University of Birmingham, Birmingham.

Arconada, Rodriguez, Santiago, 1996, "La Experiencia de Antimano. Reflexiones sobre algunos aspectos de la lucha por la Constitución de los Gobiernos Parroquiales en el Municipio Libertador de la ciudad de Caracas", *Revista Venezolana de Economía y Ciencias Sociales*, Vol. 2, No. 4, pp. 155-165, Caracas.

Arconada, Rodriguez, Santiago, 2005a, "The Venezuelan Experience in the Struggle for People-Centred Drinking Water and Sanitation Services", in Belén Balanyá, Brid Brennan, Olivier Hoedeman, Satoko Kishimoto and Philipp Terhorst (eds.), *Reclaiming Public Water: Achievements, Struggles and Visions from around the World.*

Arconada, Rodriguez, Santiago, 2005b, "Seis anos después: mesas tecnicas y consejos comunitarios de agua", *Revista Venezolana de Economía y Ciencias Sociales*, Vol. 11, No. 3, pp. 187-203.

Arnstein, Sherry R., 1969, "A ladder of citizen participation", *American Institute of Planners Journal*, July, Vol. 35, No. 4, pp. 216-224.

Asamblea Nacional Constituyente, 1999, *Constitución de la República Bolivariana de Venezuela*, Gaceta Oficial Extraordinaria, No. 5453.

Asamblea Nacional de la Republica Bolivariana de Venezuela, 2001, *Ley Organica de la Contraloria General de la Republica y del Sistema Nacional de Control Fiscal*, Gaceta Oficial No. 37347, 17th of December 2001.

Asamblea Nacional de la Republica Bolivariana de Venezuela, 2001, *Ley Organica para la Prestacion de los Servicios de Agua Potable y Saneamiento*, Gaceta Oficial Extraordinaria No. 5568, 31st of December 2001.

Asamblea Nacional de la Republica Bolivariana de Venezuela, 2002, *Ley de los Consejos Locales de Planificación Pública*, Gaceta Oficial No. 37463, 12th of June 2002.

Asamblea Nacional de la Republica Bolivariana de Venezuela, 2002, *Ley del estatuto de la function publica*, Gaceta Oficial No. 37522, 6th of September 2002.

Asamblea Nacional de la Republica Bolivariana de Venezuela, 2003, *Ley contra la Corrupcion*, Gaceta Oficial No. 37522, 6th of September 2002.

Asamblea Nacional de la Republica Bolivariana de Venezuela, 2005, *Ley Organica del Poder Pùblico Municipal*, Gaceta Oficial No. 38204, 8th of June 2005; No. 38327, 2nd of December 2005.

Asamblea Nacional de la Republica Bolivariana de Venezuela, 2006, *Reforma de la Ley de los Consejos Locales de Planificación Pública*, Gaceta Oficial No. 38591, 26th of December 2006.

Asamblea Nacional de la Republica Bolivariana de Venezuela, 2007, *Ley de Reforma Parcial de la Ley Orgánica para la Prestación de los Servicios de Agua Potable y Saneamiento*, Gaceta Oficial No. 38763, 6th of December 2007.

Asamblea Nacional de la Republica Bolivariana de Venezuela, 2007, *Ley de Aguas*, Gaceta Oficial No. 38595, 2nd of January 2007.

Asamblea Nacional de la Republica Bolivariana de Venezuela, 2010, *Ley de la comunas*, 21st December 2010.

Asamblea Nacional de la Republica Bolivariana de Venezuela, 2010, *Ley de reforma parcial de la ley organica del poder publico municipal, 2010*, Gaceta Oficial No. 6015 extraordinario.

Asamblea Nacional de la Republica Bolivariana de Venezuela, 2013, *Ley del Plan de la Patria: Segundo Plan Socialista de Desarrollo Económico y Social de la Nación 2013-2019*, Gaceta Oficial de la República Bolivariana de Venezuela No. 6118 Extraordinario, 4th of December 2013.

Aubriot, Julie and Luisa Moretto, 2013, "Le droit à l'eau au Sud, un outil paradoxal pour penser le droit à la ville: entre coproduction et contestation des services urbains", in Boniburini, I., Le Maire, J., Moretto, L. and H. Smith (eds.), *The City as a common good: urban planning and the right to the city*, La Lettre Volée, Bruxelles, pp. 62-81.

Baetz, Juan Francisco, 2001, "La importancia del Régimen Economico-Financiero", in *Vertientes*, No. 11, pp. 32-34, Caracas, http://hidrocapital.gov.ve.

Bakker, Karen, 2010, *Privatizing Water: Governance Failure and the World's Urban Water Crisis*, New York: Cornell University Press, Ithaca.

Balbo, Marcello, 2002a, "Dalla good governance all'urban governance. Il percorso urbano della Banca Mondiale", *Urbanistica*, No. 118, pp. 13-19.

Balbo, Marcello, 2002b, "La gestione urbana decentrata", in *La città inclusiva*, Balbo Marcello (ed.), Franco Angeli, Milano.

Baron, Catherine (in collaboration with Anne Isla), 2005, "Modèles d'accès à l'eau dans les villes d'Afrique Sub-saharienne: entre efficacité et équité", *Sciences de la Société*, No. 64, pp. 157-174.

Batley, Richard and Claire Mcloughlin, 2010, "Engagement with Non-State Service Providers in Fragile States: Reconciling State-Building and Service Delivery", *Development Policy Review*, Vol. 28, No. 2, pp. 131-154.

Batley, Richard and Dominique Moran, 2004, *Literature Review of Non-State Provision of Basic Services*, International Development Department, School of Public Policy, The University of Birmingham, Birmingham.

Batley, Richard, 2006, "Engaged or divorced? Cross-service findings on government relations with non-state service-providers", *Public Administration and Development*, Vol. 26, No. 3, pp. 241-251.

Batley, Richard and George Larbi, 2004, *The Changing role of Government. The Reform of Public Services in Developing Countries*, Palgrave, Houndmills and New York.

Batley, Richard, 2004, "The Politics of Service Delivery Reform", in *Development and Change*, Vol. 35, No. 1, pp. 31-56.

Beall, Jo, 2001, "Valuing Social Resources or Capitalizing on Them? Limits to Pro-poor Urban Governance in Nine Cities of he South", *International Planning Studies*, Vol. 6, No. 4, pp. 357-375.

Benequista, Nicholas and Gaventa, John, 2011, *Blurring the Boundaries. Citizen Action Across States and Societies*, Brighton: Development Centre on Citizenship, Participation and Accountability.

Bertelemy, Jean-Claude, 2006, *Les privatisations en zone franc: synthèse des travaux du groupe de travail MINEFI / AFD.*

Boeninger, Edgardo, 1991, "Governance and Development: Issues and Constraints", *Proceedings of the World Bank Annual Conference on Development Economics 1991*, World Bank, Washington.

Booth, David, 2011a, "Governance for development in Africa: building on what Works", *Policy Brief* 01, April 2011, APPP.

Booth, David, 2011b, "Towards a theory of local governance and public goods provision", *IDS Bulletin*, Vol. 42, No. 2, pp. 11-21.

Borgucci, Emmanuel, 2003, "Representaciones y discurso en los procesos de decentralizacion administrativa en Venezuela", *Revista de Ciencias Sociales*, Vol. 13, No. 3, pp. 405-430.

Borgucci, Emmanuel, 2013, "Principales discursos que buscaban diseñar un estado descentralizado antes del surgimiento de la COPRE en Venezuela", *TELOS. Revista de Estudios Interdisciplinarios en Ciencias Sociales*, Vol. 15 (3): 429-458.

Bovaird, Tony, and Elke Loeffler, 2012, "From Engagement to Co-production: The Contribution of Users and Communities to Outcomes and Public Value", *Voluntas*, Vol. 23, No. 4, pp. 1119-1138.

Bovaird, Tony, 2007, "Beyond Engagement and Participation: User and Community Coproduction of Public Services", *Public Administration Review*, Vol. 67, No. 5, pp. 846-860.

Brewer-Carías, Allan R., 2010, *Dismantling democracy in Venezuela: the Chávez authoritarian experiment*, Cambridge University Press, New York.

Brewer-Carías, Allan R., 2011, "Las leyes del poder popular dictadas en venezuela en diciembre de 2010, para transformar el estado democrático y social de derecho en un estado comunal socialista, sin reformar la constitución", *Cuadernos Manuel Giménez Abad*, No. 1, pp. 127-131.

Briceño, Héctor, 2009, "Participación electoral y cultura política en Venezuela 1958-2007", *Cuadernos de Cendes*, No. 72, pp. 37-65.

Brillemborg, Alfredo, Kristin Feireiss and Hubert Klumpner, 2005, *Informal City: Caracas Case*, Prestel, German Federal Cultural Foundation and Caracas Urban Think Tank.

Brown, John, 2001, *De la gouvernance ou de la constitution politique du néo-libéralisme* (website: www.france.attac.org/a832).

Brudney, Jeffrey L. and Robert E. England, 1983, "Toward a Definition of the Coproduction Concept", *Public Administration Review*, Vol. 43 (January/ February), pp. 59-65.

Burgess, Rod, 2002, "Sviluppo Urbano Sostenibile", in *La città inclusiva*, Balbo Marcello (ed.), Franco Angeli, Milano.

Burgess, Rod, Marisa Carmona and Theo Kolstee (eds.), 1997, *The Challenge of Sustainable Cities*, Zed Books Ltd, London.

Canache, Damarys, 2004, "Urban Poor and Political Order", in McCoy, Jennifer L. and David J. Myers (eds.), *The Unravelling of Representative Democracy in Venezuela*, The Johns Hopkins University Press, Baltimore and London.

Cariola, Cecilia and Miguel Lacabana, 2003, *WSS Practices and Living Conditions in the Peri-Urban Interface of Metropolitan Caracas: the case of Bachaquero and Paso Real*, Development Planning Unit, UCL, London.

Cariola, Cecilia and Miguel Lacabana, 2005a, "Los bordes de la esperanza: nuevas formas de participación popular y gobiernos locales en la periferia de Caracas", *Revista Venezolana de Economía y Ciencias Sociales*. Vol. 11, No. 1, pp. 21-42, Caracas.

Cariola, Cecilia and Miguel Lacabana, 2005b, "Construyendo la participacion popular y una nueva cultura del agua en Venezuela", *Cuadernos del Cendes*, Vol. 22, No. 59, pp. 111-136, Caracas.

Cariola, Cecilia and Miguel Lacabana, 2007, *Entre la ciudad global y la periferia en transicion: Caracas – Valles del Tuy Medio*, Ministerio del Poder Popular para Ciencia y Tecnologia, Caracas.

Castellanos Vasquez, Jesùs, 2005, *Elecciones Municipales Venezolanas 2005: Un proceso de dificil analisis*, Setie "Documentos de Trabajo", Observatorio Electoral Latinoamericano, www.observatotioelectoral.org.

Castro, Esteban, 2005, "Agua y gobernabilidad: entre la ideologia neoliberal y la memoria histotica", *Cuadernos del CENDES*, Vol. 22, No. 59, pp. 3-22, Caracas.

Castro, Esteban, 2006, *Water, Power, and Citizenship. Social Struggle in the Basin of Mexico*, Palgrave-Macmillan.

Cavill, Sue and M. Sohail, 2004, "Strengthening accountability for urban services", *Environment & Urbanization*, Vol. 16, No. 1, pp. 155-170.

Chavez, Daniel and Benjamin Goldfrank (eds.), 2004, *The left in the City. Participatory Local Governments in Latin America*, Latin American Bureau, London.

Coase, Ronald, 1937, "The Nature of the firm", *Economica*, New Series, Vol. 4, No. 16, pp. 386-405.

Coing, Henry, 1995, "Changer les règles du jeu: Caracas, chronique d'une privatisation manquée", in D. Lorrain (dir.), *Gestions urbaines de l'eau*, Economica, Paris, pp. 99-115.

Collignon, Bernard and Marc Vézina, 2000, *Independent Water and Sanitation Providers in African Cities. Full Report of a Ten-Country Study*, Water and Sanitation Program, WBI, GTZ, BADC.

Committee on Economic, Social and Cultural Rights, 1966, *International Covenant on Economic, Social and Cultural Rights*.

Committee on Economic, Social and Cultural Rights, 2002, *General Commitment No. 15*.

Córdova, Jaimes, 2012, "Los desafíos de la política comunitaria en Venezuela", *Revista de Ciencias Sociales*, Vol. XVIII, No. 4, pp. 670-683.

Corrales, Maria Elena, 1998, *El reto del Agua. Cambios institutionales en los servicios publicos por redes*, Galac, Caracas.

Corrales, Maria Elena, 2002, *Gobernabilididad de los servicios de agua potable y saneamiento en America Latina*, SAMTAC – GWP, Caracas, Venezuela.

Corrales, Maria Elena, 2004, *Venezuela: Analisis del Sector Agua Potable y Saneamiento*, Corporacion Andina de Fomento – CAF, Vicepresidencia de Infraestructura, Informe Sectoriales de Infraestructura, año 2, No. 2, marzo de 2004.

COWI, 2005, *Examination of Potential of the ACP-EU Water Facility for Encouraging Increasing and Innovative Financing in Water and Sanitation*, Final Report.

Crane, Randall and Amrita Daniere, 1996, "Measuring access to basic services in global cities", *Journal of the American Planning Association*, 0194-4363, Vol. 62, issue 2, pp. 203-221.

Cuñarro, Edith, Eduviges Morales and Nila Leal Gonzales, 2006, "Nueva Gerencia Publica en la Constitucion de la Republica Bolivariana de Venezuela", *Revista Venezuelana de Gerecia*, año 11, No. 33, pp. 122-132.

Cunill, Nuria, 2004, "La democratizacion de la Administracion Publica. Los mitos a vencer", *Politica y Gestion Publica*, Editorial Fondo de Cultura CLAD, pp. 43-90, Argentina.

D'Elia, Yolanda, 2010, *En Defensa de la Descentralización en Venezuela*, working paper.

Daniere, Amrita G. and Lois M. Takahashi, 1999, "Poverty and Access: Differences and Commonalties across Slum Communities in Bangkok", *Habitat International*, Vol. 23, No. 2, pp. 271-288.

De la Cruz, Rafael, 2004, "Decentralisation: Key to Understanding a Changing Nation", in McCoy, Jennifer L. and David J. Myers (eds.), *The Unravelling of Representative Democracy in Venezuela*, The Johns Hopkins University Press, Baltimore and London.

De Lisio, Antonio, 2009, "La metrópoli caraqueña: expansión simplificadora o articulación diversificante", *Terra*, Vol. XXV, No. 38, pp. 191-226.

Degnbol-Martinussen, John and Poul Engberg-Pedersen, 2003, *Aid: Understanding International Development Cooperation*, Zed Books Ltd, London.

Delgado, José Gregorio, 2004, *Transparencia, Contraloria Social y Participacion ciudadana en los Asuntos Publicos: Leyes para su Defensa*, www.worldbank.org.

Derham, Michael, 2002, "Undemocratic Democracy: Venezuela and the Distorting of History", *Bulletin of Latin American Research*, Vol. 21, No. 2, pp. 270-289.

Devas, Nick, 1999, *Who runs cities? The Relationship between Urban Governance, service Delivery and Poverty*, Theme Paper 4, IDD, School of the Public Policy, University of Birmingham.

Devas, Nick, 2001, "Does City governance Matters for the Urban Poor?", *International Planning Studies*, Vol. 6, No. 4, pp. 393-408.

Didier, Sophie, Marianne Morange and Elisabeth Peyroux, 2012, "City Improvement Districts and 'Territorialised Neoliberalism' in South Africa (Johannesburg, Cape Town)", in Künkel J. and M. Mayer (eds.), *Neoliberal urbanism and its contestations: crossing theoretical boundaries*, Houndmills, Basingstoke Hampshire, Palgrave Macmillan, New York.

Dietz, Henry A. and David J. Myers (eds.), 2002, *Capital City Politics in Latin America*, Lynne Rienner Publishers, London.

Ellner, Steve and David J. Myers, 2002, "Caracas: Incomplete Empowerment Amid Geopolitical Feudalism", in Henry A. Dietz and Myers David J. (eds.), *Capital City Politics in Latin America*, Lynne Rienner Publishers, London.

Englebert, Pierre, 2001, "The World Bank and the Unsuspected Virtue of 'Social Capital', unedited English version of 'La Banque Mondiale et les vertus insoupçonnées du 'capital social'", in Gilbert Rist (ed.), *Les mots du pouvoir: sens et non-sens de la recherche internationale*, Presses universitaires de France, Paris.

Environment & Urbanization, 2000, *UNCHS (Habitat) – the global campaign for good urban governance*, Vol. 12, No. 1, pp. 197-202.

Erzberger, Cristian and Udo Kelle, 2003, "Making inferences in mixed methods: the rules of integration", in *Handbook of Mixed Methods in Social and behavioural Research*, Abbas Tashakkori and Charles Teddlie editors, SAGE Publications, Thousands Oaks, California.

Escalona, Lenny, Manuel Espita and Lucia Garcia, 2009, "Descripcion y caracterizacion del sector agua potable en Venezuela", *Cayapa: Revista Venezolana de Economia Social*, Vol. 9, No. 18, pp. 9-31.

Ferrer, Mercedes, Yajaira Garcia, Carolina Quintero and Helen Barroro, 2005, "Construir la ciudad. Interface entre politica urbana y finanzas publicas locales", *Revista de Ciencias Sociales*, Vol. 11, No. 3, pp. 507-528.

Figueroa, Vistor M., 2006, "The Bolivarian Government of Hugo Chavez: Democratic Alternative for Latin America?", *Critical Sociology*, Vol. 32, issue 1, pp. 187-211.

Francisco, Cristobal, 2005, "Cambio y equidad nel servicio de agua en Venezuela", *Cuadernos del Cendes*. Vol. 59, No. 3, pp. 145-159, Caracas.

Francisco, Cristobal, 2006, *Sector Agua Potable y Saneamiento en Venezuela (1998-2005)*, Hidroven presentation.

Garcia, Marisol, 2006, "Citizenship, Practices and Urban Governance in European Cities", *Urban Studies*, Vol. 43, No. 4, pp. 745-765.

Gaventa, John and Gregory Barrett, 2010, "So What Difference Does it Make? Mapping the Outcomes of Citizen Engagement", *IDS Working Paper 347.*

Gaventa, John, 2004, "Towards participative governance: assessing the transformative possibilities", in Hichey Samuel and Giles Mohan (eds.), *Participation: From Tyranny to Transformation? Exploring New Approaches to Participation in Development*, Zedbooks, New York.

Gibbs, Terry, 2006, "Business as Unusual: what the Chavez era tells us about democracy under globalisation", *Third World Quarterly*, Vol. 27, No. 2, pp. 265-279.

Gil Yepes, José Antonio, 2004, "Public Opinion, Political Socialisation, and Regime Stabilisation", in McCoy, Jennifer L. and David J. Myers (eds.), *The Unravelling of Representative Democracy in Venezuela*, The Johns Hopkins University Press, Baltimore and London.

Goldfrank, Benjamin, 2001, *Deepening Democracy through Citizen Participation? A comparative Analysis of Three Cities*, prepared for the delivery at the 2001 meeting of the Latin American Studies Association, Washington DC, September 6-8, 2001.

Goldfrank, Benjamin, 2004, "Libertador: La Causa R Meets Caracas the Horrible", in Daniel Chavez and Benjamin Goldfrank (eds.), *The left in the City. Participatory Local Governments in Latin America*, Latin American Bureau, London.

Goldfrank, Benjamin, 2011, "Los Consejos Comunales: ¿Avance o retroceso para la democracia venezolana?", *Íconos. Revista de Ciencias Sociales*, Vol. 5, No. 39, pp. 41-55.

Gonzales Landazabal, Angela B., 2000, *Agua para el Siglo XXI para America del Sur: De la Vision a la Accion. Informe nacional sobre la gestion del agua en Venezuela*, Global Water Partnership and Cepal.

González Villar, Carlos, 2004, *Descentralización y reforma política estatal en América Latina*, Posadas, Misiones: Editorial Universitaria, Universidad Nacional de Misiones.

Guarneroz-Meza, Valeria and Mike Geddes, 2010, "Local Governance and Participation under Neoliberalism: Comparative Perspectives", *International Journal of Urban and Regional Research*, Vol. 34, No. 1, pp. 115-129.

Halfani, Mohamed, Patricia McCarney and Alfredo Rodriguez, 1995, "Towards an Understanding of Governance", in Richard Stren with Judith Bell (eds.), *Urban Research in the Developing World Vol. 4*, Perspective on the City (Toronto: Centre for Urban and Community Studies, University of Toronto) pp. 91-14.

Harpham, Trudy and Kwasi Boateng, 1997, "Urban Governance in Relation to the Operation of Urban Services in Developing Countries", *Habitat International*, Vol. 21, No. 1, pp. 65-77.

Harvey, David, 2005, *A brief history of neoliberalism*, Oxford University Press.

Harvey, David, 2012, *Rebel Cities: From the Right to the City to the Urban Revolution*, Verso Books.

Hellinger, Daniel, 2005, "When 'No' Means 'Yes to Revolution'", *Latin American Perspectives*, issue 142, Vol. 32, pp. 8-32.

Hernaiz, Carlos, 2004, *La Participacion Ciudadana en las Mesas Tecnicas de Agua (Estudio de Caso del sector Las Margaritas de la parroquia La Vega)*, Trabajo Final de Grado, Universidad Central de Venezuela, Facultad de Ciencias Economicas y Sociales, Escuela de Sociologia, Caracas.

Hichey, Samuel and Giles Mohan (eds.), 2004, *Participation: From Tyranny to Transformation? Exploring New Approaches to Participation in Development*, Zedbooks, New York.

Hickey, Samuel, 2010, "The Government of Chronic Poverty: From Exclusion to Citizenship?", *The Journal of Development Studies*, Vol. 46, No. 7, pp. 1139-1155.

Hidrocapital, 2005, *Frecuencia de servicio de agua potable del sistema Losada-Ocumarito*, mimeo, power point presentation.

Hidroven and MARN, 2005, *El agua que nos une: Informe de Gestion del sector Agua Potable y Saneamiento*, Caracas.

Hidroven, 2005, Venezuela, "III Encuentro de experiencias comunitarias en agua potable y saneamiento", *Quadernos del CENDES*, año 22, No. 59, pp. 173-174, Caracas.

Higuerey Ángel, María Manuela González and Lourdes Trujillo, 2012, "Situación actual de los servicios de suministro de agua en Venezuela", in *Cayapa: Revista Venezolana de Economia Social*, Vol. 12, No. 4, pp. 9-36.

Hydén, Göran and John Samuel (eds.), 2011, *Making the State Responsive: Experience with Democratic Governance Assessments*, United Nations Development programme.

Hydén, Göran, 2011, "Making the State Responsive: Rethinking Governance Theory and Practice", in Hydén, Göran and John Samuel (eds.), *Making the State Responsive: Experience with Democratic Governance Assessments*, United Nations Development programme, pp. 5-28.

IMUTC (2012). Avances del Plan Estratégico Caracas Metropolitana 2020. Alcaldía Metropolitana de Caracas. Caracas.

INDIA – Instituto de Investigation y Defensa Integral Autogestionaria, 2006, "Informe sobre la Observancia del Derecho al Agua en venezuela", in *Agua: Constitution Social de un Derecho Umano.*

INE – Instituto Nacional de Estadistica, 2001, *Censo 2001*, www.ine.gov.ve.

INE – Instituto Nacional de Estadistica, 2011, *XIV Cenco National de Poblacion y Vivienda*, www.ine.gov.ve.

Inter-American Development Bank, 1998, *Program to Support Decentralisation of the Potable Water and Sanitation Sector*, www.iadb.org.

Inter-American Development Bank, 2011, *Water Consumption Efficiency Program (PRAC).*

Jaglin, Sylvy, 2001, "L'eau potable dans les villes en développement: les modèles marchands face à la pauvreté", *Revue Tiers Monde*, t. XLII, No. 166, pp. 275-303.

Jakobsen, Morten, 2012, "Can Government Initiatives Increase Citizen Coproduction? Results of a Randomized Field Experiment", *Journal of Public Administration Theory and Practice*, Vol. 23, No. 1, pp. 27-54.

Jessop, Bob, 2001, *Good Governance and the Urban Question: On Managing the Contradictions of Neo-Liberalism*, published by the Department of Sociology, Lancaster University, Lancaster LA1 4YN, http://www.comp.lancs.ac.uk/sociology/papers/Jessop-Good-Governance-and.the-Urban-Question.pdf.

Jessop, Bob, 2002, "Liberalism, Neoliberalism, and Urban Governance: A State-Theoretical Perspective", *Antipode*, Blackwell Publishers, pp. 452-472.

Jiménez, Raul and Manuel Hidalgo (2012), "Forensic Analysis of Venezuelan Elections during the Chavez Presidency", *Plos One*, Vol. 9, No. 6.

Joint Monitoring Programme for Water Supply and Sanitation, 2014, *Venezuela (Bolivarian Republic of): estimates on the use of water sources and sanitation facilities (1980-2012)*, WHO/UNICEF, http://www.wssinfo.org/.

Joshi, Anuradha and Mick Moore, 2004, "Institutionalised Co-production: Unorthodox Public Service Delivery in Challenging Environments", *The Journal of Development Studies*, Vol. 40, No. 4, pp. 31-49.

Kariuki, Mukami and Jordan Schwartz, 2005, *Small-Scale Private Service Providers of Water Supply and Electricity: A Review of Incidence, Structure, Pricing and Operating Characteristics*, World Bank Policy Research Working Paper 3727.

Kaufmann, Daniel, Frannie Léautier and Massimo Mastruzzi, 2004, *Governance and the City: An Empirical Exploration into Global Determinants of Urban Performance*, World Bank Institute Discussion Paper.

Kjellén, Marianne and Gordon McGranaham, 2006, *Informal Water Vendors and the Urban Poor*, Human Settlements Discussion Paper Series, Theme: Water.

Knoop, Joachim, 1994, *La descentralizacion Fiscal en Venezuela*, Proyect Regional de Descentralizacion Fiscal, Cepal-GTZ, Santiago (Chile).

Kooiman, Jan, 2003, *Governing as Governance*, Sage, London.

Kundu, Amitabh, 2003, "Urbanisation and Urban Governance: Search for a Perspective beyond Neo-Liberalism", *Economic and Political Weekly*, 38 (29).

Künkel, Jenny and Margit Mayer (eds.), 2012, *Neoliberal urbanism and its contestations: crossing theoretical boundaries*, Houndmills, Basingstoke Hampshire, Palgrave Macmillan, New York.

La Voz, 2005, Valles del Tuy, *En Paz Castillo está asegurado triunfo absoluto de las morochas*, 05/07/2005.

Lacabana, Miguel (coord.), Cecilia, Cariola and Veronica, Carrodeguas, 2004, *El Marco Regulatorio del SAPS: Oportunidades, Restricciones y Desafios para mejorarlo. El caso de la Region Metropolitana de Caracas, Venezuela*, The Development Planning Unit, University College London.

Lacabana, Miguel (in collaboration with Cecilia Cariola, Rosaura Sierra, Marianela Carillo), 2003, *An Overview of the Water Supply and Sanitation System at Metropolitan and Peri-urban Level: the case of Caracas*, Development Planning Unit, UCL, London.

Lacabana, Miguel, Cecilia Cariola, Irama La Rosa, Alejandro Maldonado Fermín, Magally Villasmil, Lupita Vielma, 2008, *Las Mesas Técnicas de Agua en el Contexto de los Cambios Institucionales, la Democracia Participativa y la Participación Popular en Venezuela*, Informe final del Proyecto: "Mesas Técnicas de Agua, proyectos comunitarios y análisis de los procesos de participación", Contrato No. GPFFC-02-2007 (CENDES-HIDROVEN).

Lalander, Rickard and Francisco Roberto Garcia Samaniego, 2005, "Chavismo y oposicion en Venezuela: Exploraciones criticas sobre democracia, descentralizacion y populismo", *Ciudad Politica. Praxis y Ciencia Politica*, http://www.ciudadpolitica.com.

Lander, Edgardo, 2004, *Venezuela: la busqueda de un proyecto contrahegemonico*, www.cadtm.org.

Laws, Sophie, Caroline Harper and Rachel Marcus, 2003, *Research for Development*, Sage Publications and Save the Children.

Leal Rios, Gleccy and Juan Carlos Gotera, 2003, "Politica de descentralizacion fiscal hacia los municipios en Venezuela: ¿un arreglo institucional posible?", *Ciencia de Gobierno*, Vol. 7, No. 13, pp. 123-145, Instituto Zuliano de Estudios Politicos, Economicos y Socilaes (IZEPES), Maracaibo, Venezuela.

Leftwich, Adrian, 1994, "Governance, the State and the Politics of Development", *Development and Changes*, Vol. 25, No. 2, pp. 363-386, Oxford.

Leitner, Helga, Jamie Peck and Eric S. Sheppard, 2007, *Contesting neoliberalism: Urban Frontiers*, New York, Guilford Press.

Levine, Daniel H. and Catalina Romero, 2004, "Movimientos Urbanos y Desempoderamiento en Perù y Venezuela", *America Latina Hoy*, 36, pp. 47-77, Universidad de Salamanca, Salamanca, España.

Levine, Daniel H. and Catalina Romero, 2006, *Urban Citizen Movements and Disempowerment in Peru and Venezuela*, mimeo, www.ereserves.nd.edu.

Levine, Daniel H., 2002, "The Decline and Fall of Democracy in Venezuela: Ten Theses", *Bulletin of Latin American Research*, Vol. 21, No. 2, pp. 248-269.

Lewis, David, 2003, *Civil Society in Non-Western Contexts: Some Reflections on the "Usefulness" of a Concept*, Summer School in Analysis and Governance for Sustainable Development, Venice.

Lombard, Melanie, 2013, "Citizen Participation in Urban Governance in the Context of Democratization: Evidence from Low-Income Neighbourhoods in Mexico", *International Journal of Urban and Regional Research*, Vol. 37, No. 1, pp. 135-150.

Lopez Maya, Margarita, 2008, "Innovaciones participativas en la Caracas bolivariana: la MTA de la pedrera y la OCA de barrio Unión-Carpintero", *Revista Venezolana de Economía y Ciencias Sociales*, Universidad Central de Venezuela, Venezuela, Vol. 124, No. 1, pp. 65-93.

Lopez Maya, Margarita and Lander, Luis E. 2011, "Participatory Democracy in Venezuela: Origins, Ideas, and Implementation", in Smilde, David and Daniel Hellinger (eds.), *Venezuela's Bolivarian Democracy: Participation, Politics, and Culture under Chávez*, Duke University Press, pp. 58-79.

Lopez Vallador, Mirtha Carmen Añes Hernandez, 2005, "Los Consejos Locales de Planificacion Publica en Venezuela: Una Instancia de Participacion y Capacitacion Ciudadana", *Revista Venezuelana de Economia Social*, año 5, No. 10, pp. 114-141.

Lyons, Michal, Carin Smuts and Andrea Stephens, 2001, "Participation, Empowerment and Sustainability: (How) Do the Links Work?", *Urban Studies*, Vol. 38, No. 8, pp. 1233-1251.

Maingon, Thais, 2009, "Ciudadania y Elecciones en Venezuela: ¿campos en tension?", *21st World Congress of Political Science*, Santiago de Chile, 12-16 julio 2009.

Maloutas, Thomas and Maro Pantelidou Malouta, 2004, "The Glass Managerie of Urban Governance and Social Cohesion: Concepts and Stakes / Concepts as Stakes", *International Journal of Urban and Regional Research*, Vol. 28, No. 2, pp. 449-465, Oxford.

Manchotte, Marie-Thérèse, 2000, "The European Commission's urban programmes for developing countries", *Villes en Développement*, No. 48, pp. 5-6.

Manchotte, Marie-Thérèse, 2001, *Les interventions européennes dans le domaine urbain*, mimeo, Bruxelles.

Manchotte, Marie-Thérèse, Payne Goffrey et Act consultants, 1997, *Evaluation des projets de développement urbain, Phase de synthèse et d'internalisation*, Bruxelles.

Marcano, Ester Elena, 1993, *La Crisis del Agua en Caracas: Elementos para el Analisis de la Politica Urbana*, Universidad Central de Venezuela, Consejo de Desarrollo Cientifico y Humanistico, Caracas.

Marin, Philippe, 2009, *Public-Private Partnerships for Urban Water Utilities: A Review of Experience in Developing Countries*, World Bank/PPIAF, Trends and Policy Options, No. 8.

Martínez, Roger, 2013, *La gestion del agua potable y el saneamiento en el Area Metropolitana de Caracas*, ildis, Instituto Latinoamericano de Investigaciones Sociales.

Mascareño, Carlos, 1996, *Municipalizacion de los Servicios Sociales en America Latina*, CLAD and AECI.

Mascareño, Carlos, 2003, "La descentralizacion en Venezuela", *Proces de decentralizacion en la Comunidad Andina*, Ed. Fernando Camon.

Mayer, Margit and Jenny Künkel, 2012, "Introduction: Neoliberal Urbanism and its Contestations – Crossing Theoretical Boundaries", in Künkel J. and M. Mayer (eds.), *Neoliberal urbanism and its contestations: crossing theoretical boundaries*, Houndmills, Basingstoke Hampshire, Palgrave Macmillan, New York.

McCoy, Jennifer L., 2004, "From Representative Democracy to Participatory Democracy? Regime Transformation in Venezuela", in McCoy, Jennifer L. and David J. Myers (eds.), *The Unravelling of Representative Democracy in Venezuela*, The Johns Hopkins University Press, Baltimore and London.

McCoy, Jennifer L. and David J. Myers, 2004, *The Unravelling of Representative Democracy in Venezuela*, The Johns Hopkins University Press, Baltimore and London.

McDonald, David A. and Greg Ruiters, 2012, *Alternatives to privatization: Public options for essential services in the global South*, Routledge – HSRC – LeftWord Books, South Africa.

McGranahan, Gordon, 2013, *Community-driven sanitation improvement in deprivedurbanneighbourhoods. Meetingthechallengesoflocalcollectiveaction, co-production, affordability and a trans-sectoral approach*, Research Report.

McGranahan, Gordon, Diana Mitlin and David Satterthwaite, 2008, "Land and Services for the Urban Poor in Rapidly Urbanising Countries", in Martine G., McGranahan G., Montgomery M. and Fernandez-Castilla R. (eds.), *The New Global Frontiers: Urbanisation, Poverty and Environment in the 21st Century*, Earthscan, London.

McGranahan, Gordon and David Satterthwaite, 2006, *Governance and getting the private sector to provide better water and sanitation services for the urban poor*, Human Settlements Discussion Paper Series, Theme: Water-2.

McMillan, Rebecca, Susan Spronk and Calais Caswell, 2014, "Popular participation, equity, and co-production of water and sanitation services in Caracas, Venezuela", *Water International*, Vol. 39, No. 2, pp. 201-215.

Miegville, Sophie, Thomas Ramirez, Magali Ricard, Antonin Ropert and Anne-Gaelle Wagner, 2003, *La gouvernance urbaine: un concept polysémique opératoire pour les villes des pays en développement?*, Séminaire: Problèmes économiques du développement, mimeo.

Miraftab, Faranak, 2004, "Making Neo-liberal Governance: The Disempowering Work of Empowerment", *International Planning Studies*, Vol. 9, No. 4, pp. 239-259.

Mitlin, Diana, 2008, "With and beyond the state – co-production as a route to political influence, power and transformation for grassroots organizations", *Environment and Urbanization*, Vol. 20, No. 2, pp. 339-360.

Mitlin, Diana and John Thompson, 1995, "Participatory approaches in urban areas: strengthening civil society or reinforcing the status quo?", *Environment and Urbanization*, Vol. 7, pp. 231-250.

Mitlin, Diana, 1999, "Civil Society and Urban Poverty", *Urban Governance, Partnership and Poverty*, Theme Paper 5, The University of Birmingham, Birmingham.

Mitlin, Diana, 2002, *Competition, Regulation and the Urban Poor: a Case Study of Water*, Centre on Regulation and competition, Working Paper No. 37.

Moreno, Maria Antonia, 2008, "La descentralización fiscal en Venezuela ¿subsidiariedad eficiente?", *Urbana*, No. 40, pp. 169-200.

Moretto, Luisa, 2007, "Urban Governance and Multilateral Aid Organizations: the case of Informal Water Supply Systems", *The Review of International Organizations*, Vol. 2, No. 4, pp. 345-370.

Moretto, Luisa, 2010, "Coproduction du service d'eau et recomposition de l'espace intra urbain dans la périphérie Sud de Caracas", *Espaces et sociétés*, Vol. 4, No. 143, pp. 81-99.

Moretto, Luisa, 2005, "The European Union Approach to Sustainable Urban Development in Developing Countries", Paper presented at the *11th Annual International Sustainable Development Research Conference*, June 6-8, 2005 Finlandia Hall, Helsinki, Finland.

Morse, Janice M., 2003, "Principles of mixed methods and multimethods research design", in *Handbook of Mixed Methods in Social and behavioural Research*, Abbas Tashakkori and Charles Teddlie editors, SAGE Publications, Thousands Oaks, California.

Myllylä, Susanna and Kristiina Kuvaja, 2005, "Societal premises for sustainable development in large southern cities", *Global Environmental Change*, Vol. 15, pp. 224-237.

Narang, Shipra, 2005, *Disaggregating Governance Indicators: Why Local Governance is Important and How It Can Be Measured*, paper presented at the Technical Workshop on Governance Indictors, April 20-22, New Delhi, India.

Navarro Yànez, Clemente, 2004, "Participatory Democracy and Political Opportunism: Municipal Experience in Italy and Spain (1960-93)", *International Journal of Urban and Regional Research*, Vol. 28, No. 4, pp. 819-38.

Nickson, Andrew and Richard Franceys, 2003, *Tapping the Market. The Challenge of Institutional Reform in the Urban Sector*, Palgrave, Houndmills and New York.

Nickson, Andrew, 1997, "The public-private mix in urban water supply", *International Review of Administrative Science*, Vol. 63, pp. 165-186.

Nunan, Fiona and David Satterhwaite, 2001, "The influence of Governance on the Provision of Urban Environmental Infrastructure and Service for Low-income Groups", *International Planning Studies*, Vol. 6, No. 4, pp. 409-426.

Nuñez Nava, Virginia, 2003, "La revaloracion juridica de la participacion ciudadana en la Constitucion Venezolana de 1999", *Cuestiones Politicas*, No. 30, pp. 11-25.

Oia, Ingvild, 2011, "Squaring Governance Assessments with the Accountability Agenda", in Hydén, Göran and John Samuel (eds.), 2011, *Making the State Responsive: Experience with Democratic Governance Assessments*, United Nations Development programme, pp. 63-82.

Olivier de Sardan, Jean-Pierre, 2009, *The eight modes of local governance in West Africa*, Working Paper No. 4, APPP.

Öniş, Ziya and Fikret Şenses, 2005, "Rethinking the Emerging Post-Washington Consensus", *Development And Change*, Vol. 36, No. 2, pp. 263-290, Oxford.

Onwuegbuzie, Anthony J. and Charles Teddlie, 2003, "A framework for analysing data in mixed methods research", in *Handbook of Mixed Methods in Social and behavioural Research*, Abbas Tashakkori and Charles Teddlie editors, SAGE Publications, Thousands Oaks, California.

Osborne, Stephen P. and Kirsty Strokosch, 2013, "It takes Two to Tango? Understanding the Co-production of Public Services by Integrating the Services Management and Public Administration Perspectives", *British Journal of Management*, Vol. 24, S31-S47.

Osmont, Annik, 1995, *La Banque Mondiale et les Villes. Du développement à l'ajustement*, Parigi, Khartala.

Osmont, Annik, 1998, "La gouvernance: concept mou, politique ferme", *Les Annales de la recherche urbaine*, No. 80-81, pp. 35-41.

Osmont, Annik, 1999, "La Banca Mondiale e la gestione urbana. La ricerca dell'efficacia", *Nord e Sud*, Anno XLVI.

Osmont, Annik, 2000, "Is small so beautiful?: les ambiguïtés du paradigme localiste", in *Les annales de la recherche urbaine*, No. 86, pp. 150-153.

Osmont, Annik, 2002, "La città efficiente", in *La città inclusiva*, Balbo M. (ed.), Franco Angeli, Milano.

Ostrom, Elinor, 1996, "Crossing the Great Divide: Coproduction, Synergy, and Development", *World Development*, Vol. 24, No. 6, pp. 1073-1087.

Ostrom, Elinor, 1999, "Coping with tragedies of the Commons", *Annual Review of Political Science*, Vol. 2, pp. 493-535.

Paproski, P., 1993, "Urban Governance Systems-Another Unanalysed Abstraction?", *DPU NEWS*, Issue No. 28, February, London.

Parker Christopher and Pascal Debruyne, 2012, "Reassembling the political life of community: naturalizing neoliberalism in Amman", in Künkel J. and M. Mayer (eds.), *Neoliberal urbanism and its contestations: crossing theoretical boundaries*, Houndmills, Basingstoke Hampshire, Palgrave Macmillan, New York.

Parks, Roger B. *et al.*, 1981, "Consumers as coproducers of public services: some economic and institutional considerations", *Policy Studies Journal*, Vol. 9 (Summer), No. 1, pp. 1-11.

Peck, Jamie, Théodore, Nik, and Neil Brenner, 2013, "Neoliberal Urbanism Redux?", *International Journal of Urban and Regional Research*, Vol. 37, No. 3, pp. 1091-1099.

Penfold, Michael, 2010, "La democracia subyugada: El hiperpresidencialismo Venezolano", *Revista de Ciencia Política*, Vol. 30, No. 1, pp. 21-40.

Peraza, Andrés, 2012, "La descentralización y la competencia municipal en el periodo 1999/2011", in *TELOS. Revista de Estudios Interdisciplinarios en Ciencias Sociales*, Vol. 14, No. 3, pp. 364-381.

Peters, D.H. and V.R. Muraleedharan, 2008, "Regulating India's health services: To what end? What future?", *Social Sciences and Medicine*, Vol. 66, pp. 2133-2144.

Phumpiu, Patricia and Jan Eric Gustafsson, 2008, "When are Partnerships a Viable Tool for Development? Institutions and Partnerships for Water and Sanitation Service in Latin America", *Water Resources Management*, Vol. 23, pp. 19-38.

Pierre, Jon and Guy Peters, 2000, *Governance, Politics and the State*, MacMillan Press, London.

Pierre, Jon, 2000, *Debating Governance*, Oxford University Press.

PPIAF, 2005, *Annual Report 2005*, www adb.org.

Rachadell, Manuel, 2006, "La centralizacion del poder en Venezuela", *Provincia* No. 16, julio-diciembre 2006, pp. 199-282.

Rakodi, Carole, 2000, "'Getting the Pipe Laid s One Matter and Getting the Water Flowing through the Pipe is Another': User Views on Public-sector Urban Water Provision in Zimbabwe, Sri Lanka, Ghana and India", in *International Planning Studies*, Vol. 5, No. 3, pp. 365-391.

Rakodi, Carole, 2002, "Produzione e gestione dei servizi urbani", in *La città inclusiva*, Balbo M. (ed.), Franco Angeli, Milano.

Rakodi, Carole, 2003, "Politics and performance: the implications of emerging governance arrangements for urban management approaches and information systems", *Habitat International*, Vol. 27, pp. 523-547.

Ramirez, Ronaldo, 2002, "Il paradigma qualitativo della povertà urbana", in *La città inclusiva*, Balbo M. (ed.), Franco Angeli, Milano.

Rangel Guerrero, Christi, 2010, “Municipios, consejos comunales y democracia en Venezuela”, *Procesos Históricos*, No. 17, pp. 71-92.

República Bolivariana de Venezuela, Gobierno del Estado Bolivariano de Miranda, Direccion General de Planificacion, Direccion de Ordenaccion del Territorio, 2006, *Contenido de Segmento: Ubicacion Espacial, Organizacion Politico Administrativa del Estado Bolivariano de Miranda, Division Politico Territorial y Densidad Poblacional*, http://www.miranda.gov.ve/Territorio/Delimitacion%20Territorial.pdf.

República Bolivariana de Venezuela, Ministerio de Finanzas, 2000, *Ley de Asignaciones Económicas Especiales para los Estados y el Distrito Metropolitano de Caracas derivados de Minas e Hidrocarburos.*

República Bolivariana de Venezuela, Ministerio de Finanzas, Oficina Nacional de Presupuesto, Enero 2006, *Clasificador Presupuestario de Recursos y Egresos.*

República de Venezuela, 1993, *Régimen Tarifario para la Prestación de los Servicios de Acueductos y de Recolección, Tratamiento y Disposición de Aguas Residuales*, Gaceta Oficial No. 35161, 1st of March 1993.

Reveron Escobar, Zaira, 2003, “El proceso de descentralizacion de Venezuela, desde la perspectiva de los alcaldes electos para el periodo 2000-2004”, *America Latina Hoy*, 35, pp. 199-224, Universidad de Salamanca, Salamanca, España.

Reyes, Gabriel, 2013, “Democracia y abstención en Venezuela”, in Bisbal Marcelino (ed.), *La Política y sus tramas: Miradas desde la Venezuela del presente*, UCAB and Konrad Adenauer Stiftung, pp. 83-97.

Rios, Germán, Federico Ortega and J. Sebastián Scrofina, 2012, “Sub-national Revenue Mobilization in Latin America and Caribbean Countries: The Case of Venezuela”, *IDB Working Paper Series No. IDB-WP-300.*

Rojas Caldron, Edgar, 2000, *Potencialidades de la Descentralizacion Fiscal en Venezuela, Estudio de Caso No. 47*, Universidad de Chile, Facultad de Ciencias Físicas y Matemáticas, Departamento de Ingeniería Industrial.

Rojas Hernandez, Jesùs David, 2002, “La Contraloria Municipal en el marco de la Constitucion de la Repubblica Bolivariana de Venezuela”, *Anuario de Derecho*, No. 24, pp. 31-66.

Salamanca, Luis, 2004, “Civil Society: Late Bloomers”, in McCoy, Jennifer L. and David J. Myers (eds.), *The Unravelling of Representative Democracy in Venezuela*, The Johns Hopkins University Press, Baltimore and London.

Sandelowski, Margarete, 2000, “Combining Qualitative and Quantitative Sampling, Data Collection and Analysis Techniques in Mixed-Methods Studies”, *Research in Nursing and Health*, No. 23, pp. 246-255.

Satterthwaite, David (with Gordon McGranaham and Diana Mitlin), 2005, *Community-driven development for water and sanitation in urban areas*, Wash: Water, Sanitation and Hygiene.

Sen, Amartya, 1999, *Development as Freedom*, Oxford University Press.

Shafiei, Zahed, 2011, *Contested Understanding of Urban Governance: A Case Study of Tehran, Iran*, PhD thesis, Faculty of Humanities and Social Sciences, School of Architecture, Planning and Landscape, Newcastle University.

Smets, Henri, 2006, *The right to Water in National Legislation*, Agence Française de Développement, Research Department.

Smilde, David and Daniel Hellinger (eds.), *Venezuela's Bolivarian Democracy: Participation, Politics, and Culture under Chávez*, Duke University Press.

Smilde, David, 2011 "Introduction: Participation, Politics, and Culture – Emerging Fragments of Venezuela's Bolivarian Democracy", in Smilde, David and Daniel Hellinger (eds.), *Venezuela's Bolivarian Democracy: Participation, Politics, and Culture under Chávez*, Duke University Press, pp. 1-27.

Smith, Laïla and Susan Hanson, 2003, "Access to Water for the Urban Poor in Cape Town: Where Equity Meets Cost Recovery", *Urban Studies*, Vol. 40, No. 8, pp. 1517-1548.

Smith, Mike and Mike Beazley, 2000, "Progressive Regimes, Partnerships and the Involvement of Local Communities: A Framework for Evaluation", *Public Administration*, Vol. 78, No. 4, Winter 2000, pp. 855-878.

Snell, Suzanne, 1998, *Water and Sanitation Services for the Urban Poor Small-Scale Providers: Typology & Profiles*, UNDP – World Bank: Water and Sanitation Programme.

Solo, Tova Maria, 1998, *Competition in Water and Sanitation: The Role of Small Scale Entrepreneurs*, World Bank Viewpoint No. 165.

Solo, Tova Maria, 2003, *Independent Water Entrepreneurs in Latin America: The other private sector in water services*, World Bank.

Spronk, Susan, Carlos Crespo and Marcela Olivera, 2012, "Struggles for water justice in Latin America: Public and 'social-public' alternatives", in McDonald D. and G. Ruiters, *Alternatives to privatization: Public options for essential services in the global South*, Routledge – HSRC – LeftWord Books, South Africa, pp. 421-452.

Stewart, Kennedy, 2006, "Designing good urban governance indicators: The importance of citizens participation and its evaluation in Greater Vancouver", *Cities*, Vol. 23, No. 3, pp. 196-204.

Stoquart, 2000, "The new challenges of urbanization: UNCHS solutions", *Villes en Développement*, No. 48, pp. 3-4.

Stren, Richard, 2000, *New Approaches to Urban Governance in Latin America*, mimeo, www.idrc.org.

Stren, Richard, 2012, "Donor Assistance and Urban Service Delivery in Africa", *Working Paper No. 2012/49*, United Nations University, World Institute for Development Economics Research.

Swingedouw, Erik, 2005, "Governance Innovation and the Citizen: The Janus Face of Governance-beyond-the-State", *Urban Studies*, Vol. 42, No. 11, pp. 1991-2006.

Tashakkori, Abbas and Charles Teddlie, 1998, *Mixed Methodology: Combining Qualitative and Quantitative Approaches*, SAGE Publications, Thousands Oaks, California.

Tashakkori, Abbas and Charles Teddlie, 2003, *Handbook of Mixed Methods in Social and behavioural Research*, SAGE Publications, Thousands Oaks, California.

Tashakkori, Abbas and Charles Teddlie, 2010, *Handbook of Mixed Methods in Social and behavioural Research*, Second Edition, SAGE Publications, Thousands Oaks, California.

Théodore, Nik, Jamie Peck and Neil Brenner, 2011, "Neoliberal urbanism: cities and the rule of markets", in G. Bridge and S. Watson (eds.), *The New Blackwell companion to the city*, Wiley-Blackwell, Oxford.

Tirado, Marcos, 2006, "Una nueva forma de organización social", *Revista Ambiente*, No. 72, Octubre, Caracas.

Transparencia Venezuela, 2004, *La Oficina de Atencion al Ciudadano (OAC): Un lugar para la rendicion de cuenta*s, in collaboration with Finland Embassy and Swiss Embassy, mimeo.

Tunstall, Rebecca, 2001, "Devolution and User Participation in Public Services: How They Work and What They Do", in *Urban Studies*, Vol. 38, No. 13, pp. 2495-2514.

Tuozzo, Maria Fernanda, 2004, "World Bank, Governance Reforms and Democracy in Argentina", *Bulletin of Latin America Research*, Vol. 23, No. 1, pp. 100-118.

UNDP, 2006, *Human Development Report 2006. Beyond scarcity: Power, poverty and the global water crisis*, Palgrave Macmillan, New York.

UNDP, 2007, *Governance Indicators: A Users' Guide (2nd Edition).*

UN-HABITAT and Al-Madinah Al-Munawarah, 2010, *Expert group meeting, Localising the habitat agenda Indicators: New paradigms in developing indicators*, Final Report, 8-10 February 2010.

UN-HABITAT and the EU, 2004, *Creating a World of Sustainable Cities*, Office for Official Publications of the European Communities, Luxembourg.

UN-HABITAT Global Campaign on Urban Governance, 2001, *Minutes of Inter-Agency Meeting on the Norms of Good Urban Governance*, New York, June 2001, www.unhabitat.org/governance.

UN-HABITAT, 2012, *State of the World's Cities 2012/2013: Prosperity of Cities*, Kenya.

UN-HABITAT, 2000, *Good Urban Governance: A normative Framework*, from Conference Room Paper 6, (HS/C/PC.CRP.6), 26 February 2000, Nairobi, Kenya, www.unhabitat.org.

UN-HABITAT, 2002, *The Global Campaign for Good Urban Governance*, Concept Paper, 2nd Edition, www.unhabitat.org.

UN-HABITAT, 2003a, *Water and sanitation in the world's cities: local action for global goals*, Earthscan, London.

UN-HABITAT, 2003b, "Water and Sanitation for Cities", *Habitat Debate*, Vol. 9, No. 3.

UN-HABITAT, 2003c, *Urban Governance Index: Guidelines For Second Field Testing*, Nairobi, Kenya. www.unhabitat.org/governance.

UN-HABITAT, 2004a, *Urban Governance Index: Conceptual Foundation and Field Test Report*, Nairobi, Kenya.

UN-HABITAT, 2004b, *Urban Indicators Guidelines: Monitoring the Habitat Agenda and the Millennium Development Goals*, www.unhabitat.org/governance.

UN-HABITAT, 2004c, *Urban Governance Index: Methodology Guidelines.*

UN-HABITAT, 2004d, *Dialogue on urban services: making the private sector work for the urban poor, Thematic dialogues: Urban services*, 2nd World Urban Forum, Barcelona.

UN-HABITAT, 2006a, *Meeting Development Goals in Small Urban Centres: Water and Sanitation in the World's Cities 2006*, Earthscan, London.

UN-HABITAT, 2006b, *Water for African Cities*, www.unhabitat.org, accessed on the 22/07/06.

UN-HABITAT, ADB, 2003, *Water for Asian Cities programme*, www.unhabitat.org.

UN-HABITAT, DFID, Dpu, 2001, *Implementing the Habitat Agenda, In Search of Urban Sustainability*, Development Planning Unit, University College of London, London.

United Nations, 2000, *United Nations Millennium Declaration.*

United Nations, 2004, *World Urbanization Prospects: The 2003 Revision*, New York.

Velasquez Trujillo, Hector, 2005, "Las elecciones municipales en el Estado Miranda: Un Reflejo de la Transicion Venezuelana", *Contribuciones desde Coatepec*, Universidad Autonoma del Estado de México, año/Vol. V, No. 9, pp. 69-92.

Verschuere, Bram, Taco Brandsen and Victor Pestoff, 2012, "Co-production: The State of the Art in Research and the Future Agenda", *Voluntas*, Vol. 23, No. 4, pp. 1083-1101.

Vigola, Eran, 2002, "From Responsiveness to Collaboration: Governance, Citizens, and the Next Generation of Public Management", *Public Administration Review*, Vol. 62, No. 5, pp. 527-540.

Wagner, Sarah, 2004a, *Citizen Power and Venezuela's Local Public Planning Councils*, Participatory Democracy in Venezuela, part I, www.venezuelanalysis.com.

Wagner, Sarah, 2004b, *The Legal and Practical Basis of Citizen Power in Venezuela*, Participatory Democracy in Venezuela, part II, www.venezuelanalysis.com.

Wagner, Sarah, 2004c, *Problems and Opportunities for Citizen Power in Venezuela*, Participatory Democracy in Venezuela, part III, www.venezuelanalysis.com.

Wakefield, Elinor, 2004, *Mapping Of Donors' Policies and Approaches Towards Non-State Service Providers*, IDD, University of Birmingham, mimeo.

Walsh, Kieron, 1995, *Public Services and Market Mechanisms: competition, contracting and the new public management*, Macmillan, Basingstoke and London.

Weber, Heloise, 2004, "Reconstituting the 'Third World'? Poverty reduction and territoriality in the global politics of development", *Third World Quarterly*, Vol. 25, No. 1, pp. 187-206.

WHO and UNICEF, 2000, *Global Water Supply and Sanitation Report, 2000 Report*, World Health Organisation, UNICEF and Water Supply and Sanitation Collaborative Council.

Wild, Leni, Victoria Chambers, Maia King and Dan Harris, 2012, *Common constraints and incentive problems in service delivery*, Overseas Development Institute, Working Paper 351.

Wilde, Alexandra, 2011, "The Democratization of Governance Assessments", in Hydén, Göran and John Samuel (eds.), *Making the State Responsive: Experience with Democratic Governance Assessments*, United Nations Development programme, pp. 29-48.

Wilde, Alexandra, Narang Shipra, Marie Laberge and Luisa Moretto, 2009, *A Users' Guide to Measuring Local Governance*, UNDP, Oslo Governance Centre.

Wilpert, Gregory, 2005, "Venezuela: Participatory Democracy or Government as Usual?", *Socialism and Democracy*, Vol. 19, No. 1, issue 37, pp. 7-32, www.sdonline.org.

World Bank, 1994, *World Development Report 1994: Infrastructure for Development*, Washington D.C.

World Bank, 2000a, *Cities in Transition. World Bank Urban and Local Government Strategy*, Washington D.C.

World Bank, 2000b, *Intergovernmental Fiscal Relations and Local Financial Management program; Topic 10: Budgeting*; http://www1.wordbank.or/wbiep/decentralization/Topic10_Printer.htm.

World Bank, 2003, *World Development Report 2004: Making Services Work for Poor People*, Washington D.C.

World Bank, 2004a, *The Wold Bank Groups' Program for Water Supply and Sanitation*, Water Supply and Sanitation Sector Board, www.worldbank.org.

World Bank, 2004b, *Water Supply and Sanitation. World Bank Brief*, www.worldbank.org

World Bank, 2006, *Approaches to Private Participation in Water Services: A Toolkit*, www.worldbank.org.

Yin, Robert K., 1994, *Case study research: design and methods*, Sage Publications, London.

Zanetta, Cecilia, 2001, "The evolution of the World's Bank urban lending in Latin America: from sites and services to municipal reform and beyond", *Habitat International*, Vol. 25, pp. 513-533.

Annex 1: List of interviews

Hidroven: General Manager of Community Coordination.

Hidrocapital: Community coordinator of the Losada Ocumarito system; Ex-community coordinator of the Losada Ocumarito system; General Manager of the Losada Ocumarito system; Planning Coordinator, Department of Planning and Development, Hidrocapital.

Ministry of Ambient and Natural Resources: General Director of Environmental Education and Community Participation and ex-community coordinator of the Capital District of Caracas.

Cristóbal Rojas Municipality: Director General of the Municipality; Director of "Urban and Rural Planning"; Director of "Public Infrastructure"; Director of "Financial Planning and Budget"; Director of "Urban conservation and maintenance".

Paz Castillo Municipality; Mayor; Financial Director of the Paz Castillo Municipality; Director of "Urbanism and Infrastructure"; Person Responsible for the "Public Services" sector.

Mancomunidad de Los Valles Del Tuy Medio: Lawyer of the Fund for the Integral Development of the Middle Tuy Valleys.

Hacienda El Carmen leaders: MTA leaders: sector Araguaney Amarillo, sector Colinas, sector Nueva Republica, sector El Esfuerzo; president of neighbours association, sector Araguaney Amarillo focus group with MTA leaders (sectors Araguaney Amarillo, Colinas, Nueva Republica and El Esfuerzo).

Interviews with the neighbours of the sectors Araguaney Amarillo and Colinas: 23 residents.

Paso Real 2000 leaders: MTA leaders: Terraces 2, 8, 10 and 11; President of the Comite de Tierra Urbana, Terrazza 11; President of the neighbourhood association, Terrazza 1; Focus group with MTA leaders (Terraces 2, 8, 10 and 11).

Interviews with the neighbours in terrazza 11: 26 residents

NGOs: Provea (Programa Venezolano de Educación-Acción en Derechos Humanos), Maria Elena Rodriguez; Transparency International, Transparencia Venezuela, Executive Director, Mercedes De Freitas.

UNDP: Resident Coordinator, System of the United Nations in Venezuela, David McLachlan-Karr.

World Bank: Senior Social Specialist, María Magdalena Colmenares.

European Commission: Operation Sector of the European Commission delegation in Venezuela, Massimo Azzariti; Second Secretary of the European Commission delegation in Venezuela, Gerald Hatler.

Annex 2a: Community members' responses to interviews in the *Hacienda el Carmen* settlement

Note: question numbers correspond to the order in which the questions have been put to the community residents; the principles to which the questions belong are written in italic capital letters and in brackets. Only the simple answers to the listed questions are reported here: the residents' comments to these answers are included in the main text, apart from small additional notes presented here.

	sector Araguaney Amarillo	**sector Colinas**
Question 1: existence of improvement in access to water (*EQUITY, EFFECTIVENESS*)	- water for 1/2 a year, – before it was from tank trucks and drums 10 - water for 1/2 a year, before it was from illegal connections from Colinas 1 - water every day 2 - water once every two days 9 *note: according to the interviews in this sector, drums held 250 litres and in the beginning the price was 500 bolivares and then it went up to 1500 bolivares per drum.*	- water for 1/2 a year, before it was from tank trucks and drums 9 - water for 1/2 a year, before it was from illegal connections 3 - water every day 2 - water once every two days 10
Question 2a: knowledge of MTA / condominium (*PARTICIPATION, EQUITY, ACCOUNTABILITY*)	MTA: yes: 2 no: 9 condominium: yes: 2 no: 9 Knowledge of the person in charge of water supply and problems: 8	MTA: yes: 6 no: 6 condominium: yes: 7 no: 3 no response: 2 (neither MTA nor condominium: 2) Knowledge of the person in charge of water supply and problems: 12
Question 2b: how residents get aware of MTA / condominium (*PARTICIPATION*)	'through the meetings' 'through the invitation to the meetings by the MTA leader'	'because we are paying for water through the condominium'; 'through the payment of water'

	sector Araguaney Amarillo	sector Colinas
		'there was a meeting to present the MTA, I didn't go but I asked some friends to give me the details' 'through friends' comments' 'through the meetings'
Question 3: knowledge of importance of MTA / condominium in improving water access (*EFFECTIVENESS*)	yes: 2 I don't know about MTAs / condominiums: 9	yes: 10 I don't know about MTAs / condominiums: 2
Question 4/5: existence of relations / contact with MTA leaders (*EQUITY, PARTICIPATION*)	Contact with MTA leader: yes: 8 no: 3 I spoke about the problems before with the MTA leader: yes: 2 no: 6 no response: 3 I know that I can speak with the MTA leader if there is a problem: yes: 8 no: 1 no response: 2	Contact with MTA leader: yes: 11 no: 1 I spoke about the problems before with the MTA leader: yes: 2 no: 5 no response: 5 I know that I can speak with the MTA leader if there is a problem: yes: 8 no: 2 no response: 2
Question 6: participation in the census (*EFFECTIVENESS, PARTICIPATION*)	yes: 9 don't know 2 *note: many people remember the fact that MTA leaders asked them for their identity card number*	yes: 5 don't know 7 *note: some people do not remember because they moved there after the census was carried out*
Question 7: knowledge of MTA purpose (*EQUITY, ACCOUNTABILITY*)	yes: 2 no: 9	yes: 4 no: 6 no response: 2
Question 8a: knowledge of existence / outputs of meetings concerning water (*PARTICIPATION, ACCOUNTABILITY*)	yes: 8 no: 3	yes: 8 no: 4
Question 8b: election of leaders (*EQUITY, PARTICIPATION*)	yes: 1 no: 5 I don't remember / no response: 5	yes: / no: 4 I don't remember / no response: 8

	sector Araguaney Amarillo	**sector Colinas**
Question 9: willingness to be involved in / participation in MTA activities (*EFFECTIVENESS, PARTICIPATION, ACCOUNTABILITY*)	meetings: I participated: 5 I didn't participate: 6 meetings: I'm willing to participate: 10 I'm not willing to participate: 1 pipeline construction: yes: 3 no: 1 no response (all women): 7	meetings: I participated: 3 I didn't participate: 9 meetings: I'm willing to participate: 6 I'm not willing to participate: 1 pipeline construction: yes: 4 no: 2 no response (all women): 6
Question 10a: Agreement with water payment (*EQUITY, EFFECTIVENESS*)	I agree: 11 I do not agree: 0	I agree: 8 I do not agree: 1 I pay: 6 I do not pay: 3
Question 11: how the MTA can improve water services (*EFFECTIVENESS, EQUITY*)	'it would be better if it was every day' 'it is ok like this because the service has to be provided to other communities as well but it would be better if it was every day' '…siempre se puede mejorar!' 'more pressure'	'it is ok like this, it would be better once every two days so that people can understand that water is important' 'the MTA should educate, inform and make people aware about water use' 'to improve the quality of water because it contains sand' 'that everybody pays!' 'change pipelines before the sewerage system'

Annex 2b: Community members' responses to interviews in the *Paso Real 2000* settlement (Terraza 11)

Question 1: existence of improvement in access to water *(EQUITY, EFFECTIVENESS)*	yes, improvements: water for 6/8 months, every day, 6-7 hours per day, but not on Thursdays 9 yes, improvements: no direct water, but through a hosepipe connected to the new project 2 no improvements: more or less the same for 6 years 9 no improvements: before all day and now just from 8.30 am to 4 pm 1 water from the project, but now it is a bad service again 1 two connections from upwards and downwards 1 note: according to interviews in this sector, the pipeline construction took 2 weekends to complete; about two years before, the illegal pipelines broke down; one of the people saying that they have always had water said that before it came from the main pipe and 6 months before from the new project; for those who didn't see any difference in the current situation, water comes in small quantities and with very low pressure: "here we are not organised, when we have a certain level of service, the people stop working … they don't find an agreement"
Question 2a: knowledge of MTA *(PARTICIPATION, EQUITY, ACCOUNTABILITY)*	MTA: yes: 13 no: 10
Question 2b: how residents are made aware of MTA / condominium *(PARTICIPATION)*	"because the MTA leader of the terraza needed a computer" through the MTA terrace's leader and the census (x4) through the MTA leader through the initial presentation meeting (x4) "because I was a member of the MTA, but later I got much more involved with the CTU"
Question 3: importance of MTA / condominium in improving water access *(EFFECTIVENESS)*	yes: 7 no: 6 I don't know about MTAs / condominiums: 10
Question 4/5: existence of relations / contact with MTA leaders	Contact with MTA leader: Yes: 11 no: 12

(EQUITY, PARTICIPATION)	note: many residents know of the MTA leader in their terrace, but they don't have any means of contact when speaking about water problems. I spoke or know I can speak about water problems with a MTA leader: yes: 5[2] no: 13[3] no response: 5 [2] many residents (13 out of 23) affirming that they did not speak or that they are not willing to speak with MTA leaders, sustain that if they have a problem with water, they solve it personally or with the community's aid. [3] there are more people who remember having spoken to MTA leaders during the water survey.
Question 6: participation in the census *(EFFECTIVENESS, PARTICIPATION)*	yes: 18 don't know: 5
Question 7: knowledge of MTA purpose *(EQUITY, ACCOUNTABILITY, PARTICIPATION)*	yes: 8 no: 15
Question 8a: knowledge of the existence / outputs of meetings concerning water *(PARTICIPATION, ACCOUNTABILITY, EQUITY)*	yes: 3 no: 20
Question 8b: participation in the meeting for leaders election *(EQUITY, PARTICIPATION)*	yes: 7 no: 8 I don't know / no response: 8
Question 9: willingness to be involved in / participation in MTA activities *(EFFECTIVENESS, PARTICIPATION, ACCOUNTABILITY)*	meetings: I'm willing to participate: 16 I'm not willing to participate: 5 no response: 2 pipeline construction: yes, I did it: 11 yes, I would do it 5 no: 0 no response (all women): 8

	note: questions about participation in the meetings were not posed as it seems that these meetings no longer occur. * some women reported that their sons, brothers, husbands, … collaborate with the construction, so, figures do not add up to 23. In addition, some people have already realised the pipelines, while others are still waiting for the Hidrocapital project.
Question 10a: agreement with water payment *(EQUITY)*	I agree: 23 I do not agree: 0
Question 11: how the MTA can improve water services *(EFFECTIVENESS, EQUITY)*	Among those already benefiting directly from the Hidrocapital intervention: no improvements 'that all the community could benefit from the service' 'it is a good service' 'improving the pipelines because they are provisional' 'it is always possible to improve … maybe with the homes programme many problems will be solved' 'a work of awareness within the community … below, people offer water to neighbours and as a result we lack water; eventually, those who paid and worked for the project are not accessing water' (Jakelin) 'there are too many households connected to the same pipeline' 'to do the project in all the sectors' 'to change the pipes' 'the MTA could do the same work in all the terraces and in the entire settlement of PR2' 'I'm happy with it but it can always be improved by putting in new pipelines' Among those accessing water by illegal connections: '… it is always possible to improve but it is ok like it is' 'if Hidrocapital could put in the pipelines to have a better and regular service' 'I hope that something will be done because I don't have anything' 'if we accessed water during the day and night' 'through the legalisation of the service' 'to do all the work and not just half the project' 'it can be improved and with the homes programme many things can be improved' 'to change the pipes and to put in a meter' 'when they are going to do the homes programme they can also improve the water service' 'to put in a meter'

Annex 3: Calculation of the UGI indicators in the Paz Castillo and Cristóbal Rojas Municipalities

Note 1: many indicators – especially those related to water provisions – are considered for both the Paz Castillo and Cristóbal Rojas municipalities and the state-owned water company Hidrocapital. Nevertheless, the final assignment of results for these indicators has been made on the basis of municipal data.

Note 2: UGI is calculated for the year 2006, while the main text provides – when applicable and relevant – updates to the situation in that year.

EFFECTIVENESS INDICATORS:

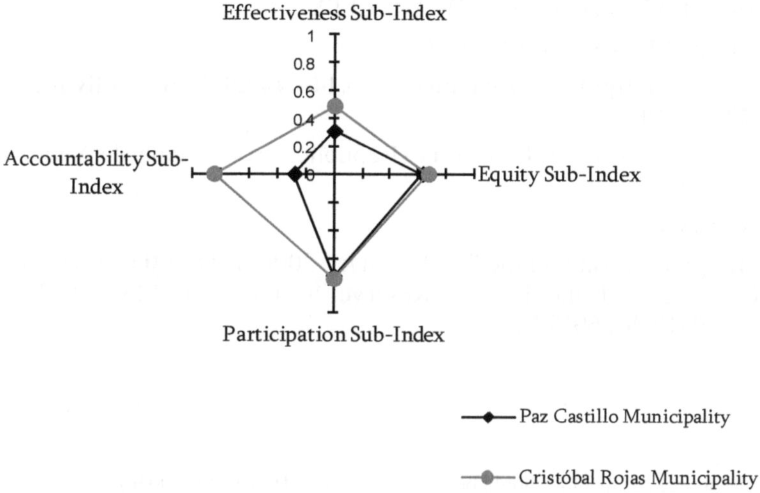

Indicator 1

Local government revenue per capita – LGR ($)

This indicator represents "the level of income which municipalities are able to raise from their residents, the business and the industries and from higher level of government" (UN-HABITAT, 2004b: 49). It is defined as the "total local government revenue from all sources in US dollars annually, both capital and recurrent for all local governments in the metropolitan area, averaged over the last three years, divided by the population" (ibid.). 'The balance between the sources of income provides an indication of the viability, independence and control over resources of the local government, and thus its effectiveness' (UN-HABITAT, 2003c).

There are two primary sources of revenue (World Bank, 2000b):

1. *revenues collected directly by the local authority (numbers 1, 2 and 3)*[1]
2. *revenues including central government transfers and grants (numbers 4, 5 and 6, see note 2).*

PAZ CASTILLO MUNICIPALITY: Estimation for the year 2006:

>>> data: 86.08 $ – result: 0.57

Local municipality revenues: 19,999,900,728.00 bolivares → 9,325,702.1 $

Population: 108,352 inhabitants (Republica Bolivariana de Venezuela *et al.*, 2006)

LGR: 86.08 $

CRISTÓBAL ROJAS MUNICIPALITY:

>>> data: 123.9 $ – result: 0.63

Local municipality revenues: 25,140,444,917.80 bolivares → 11,722,673.19 $

Population: 94,615 inhabitants (Republica Bolivariana de Venezuela *et al.*, 2006)

LGR: 123.9 $

(exchange rate: rate on the 9th of January 2006: **2,144.60** (source: Board of Governors of the Federal Reserve System, http://132.200.33.130/releases/H10/20060109/)).

---

1 The sources of income should be recorded as (UN-HABITAT, 2003c):

- taxes: municipal rates and levies, any local taxes on the transfer of property and any other taxes such as entertainment or hotel taxes, motor vehicle taxes, taxes on business which do not reflect the direct provision of services (a. property tax; b. local income tax; c. taxes on expenditures);
- user charges: include any local government charges for services provided, such as water, refuse collection, building permits and betterments levies;
- other own source income: includes interest and principal received, sales of capital items, but not donations, voluntary contributions or aid;
- transfer: formula driven payments (such as repatriation or income tax) or other grants donations from national or state government;
- loans: borrowing from all sources including bonds;
- other income sources: any other income such as donations or aid.

Indicator 2

Ratio of mandated recurrent to capital budget – RRC (%)

"The indicator ... assesses the existing distribution of local government budget sources that has direct implications on the financial sustainability"[2] *(UN-HABITAT, 2004a: 31).*

Data collection was achieved considering the mandated – instead of the actual – recurrent and capital budget. The budget items of "Income budget" estimation for the year 2006 have been separated into "recurrent budget" and "capital budget" according to the CLASIFICADOR PRESUPUESTARIO DE RECURSOS Y EGRESOS (República Bolivariana de Venezuela, Ministerio de Finanzas, Oficina Nacional de Presupuesto; Enero 2006): this separation has been realised on the base of the denomination on budget incomes and not on the base of codes (because the latter did not correspond to those obtained from the municipality).

Recurrent budget: indirect taxation, tax income, incomes from public administration goods and services selling, propriety incomes.

Capital budget: current and capital transfers and donations (income received to fund current or capital expenditures 3.05.00.00.00, República Bolivariana de Venezuela, Ministerio de Finanzas, Oficina Nacional de Presupuesto; Enero 2006).

PAZ CASTILLO MUNICIPALITY: Estimation for the year 2006:

>>> data: 0.16 – result: 0.13

Recurrent budget: 2,810,878,239.00 bolivares

Capital budget: 17,189,022,489.00 bolivares

RRC: 0.16

CRISTÓBAL ROJAS MUNICIPALITY: Estimation for the year 2006:

>>> data: 1.25 – result: 0.58

Recurrent budget: 13,975,916,115.85 bolivares

Capital budget: 11,164,528,801.95 bolivares

RRC: 1.25

Indicator 3

Percentage of local government revenue in transfers – LGT (%)

"This indicator addresses viability and independence over financial resources" (UN-HABITAT, 2004a: 32). "The transfers include the

2 As for the previous indicator, the sources of income should be recorded as: taxes, user charges, other own source income, transfer, loans, and other income sources.

formula driven payments (such as repatriation of income tax) and other grant donations from higher government levels including national or state governments" (UN-HABITAT, 2003c: 7). It is "recommended to incorporate appropriate range for local government revenue to take care of unexpected reporting" (ibid.). (n.b.: The percentage of local government revenue in transfers is generally regulated by three normative sources: Situado Constitutional; Situado Municipal; Organic decentralisation and transfer of power act).

Percentage transfer	score
0-25	*1.00*
25-50	*0.75*
50-75	*0.50*
75-100	*0.25*

PAZ CASTILLO MUNICIPALITY:

>>> data: 86% – result: 0.25

Local municipality revenues (R) (estimation 2006): **19,999,900,728.00 bolivares**

Transfer in local government revenue (T) (estimation 2006): **17,189,022,489.00 bolivares**

Percentage of local government transfer (LGT) = (T/R)*100: **86%**

Scoring of the percentage of transfers: 0.25

CRISTÓBAL ROJAS MUNICIPALITY:

>>> data: 44.4% – result: 0.75

Local municipality revenues (R) (estimation 2006): **25,140,444,917.80 bolivares**

Transfer in local government revenue (T) (estimation 2006): **11,164,528,801.95 bolivares**

Percentage of local government transfer (LGT) = (T/R)*100: **44.4%**

Scoring of the percentage of transfers: 0.75

Indicator 4

Ratio of mandate to actual tax collection – TC (%)

>>> not available

"The indicator addresses effectiveness in financial management and to some extent is a proxy to 'willingness of citizens to pay taxes'" (UN-HABITAT, 2004a: 32).

PAZ CASTILLO MUNICIPALITY:

Public officials interviewed in the Paz Castillo municipality did not know the figures concerning mandate tax and actual tax collection.

CRISTÓBAL ROJAS MUNICIPALITY:

Public officials interviewed in the Cristóbal Rojas municipality estimated that the municipality collected 50% of local taxes (mandate tax collection) but this figure was not supported by any factual data. In addition there were no figures for the actual tax collection.

Indicator 5

Predictability of transfers in local government budget – PoT (yes/no)

"The indicator addresses quality of the relevant institutions by measuring whether the procedures exist that enables the government to know the funds to be transferred in advance (intergovernmental fiscal transfers)" (un-habitat, 2004a: 33). "The indicator signifies the level of commitment and interventions by the higher level of government in local administration. ... Transfers from central or state to local government usually have some basis (e.g. Population size, taxes collected, number of social cases). One cannot predict the future transfer unless these 'basis' are stable and thus predictable. It is this basis for the amount of transfers in the past that provides an important indicator of predictability of transfers" (Un-habitat, 2003c: 7).

Thus, this indicator is composed by two different parts:

- *Is the amount to be transferred known in advance?*
- *Is there a basis for the transfer?*

PAZ CASTILLO AND CRISTÓBAL ROJAS MUNICIPALITIES:

>>> data: yes – result: 1

Public officials in both municipalities affirmed that the percentage of transfers that their municipalities received was proportional to the number of inhabitants and that detailed information was available in the *Ley Organica del Poder Pùblico Municipal* (Gaceta Oficial No. 38204, 8th of June 2005). According to this law (art. 141), municipalities can benefit from the *Situado Constitucional* each year, in proportion to 20% of the state budget belonging to recurrent incomes from the National Treasury, and at least 20% of recurrent incomes of the same state (Situado Municipal). The basis for the transfer of the *Situado Constitucional* among the municipalities of each state is:

- 30% in equal parts for each municipality;
- 50% in proportion to the municipality's population;
- 20% in proportion to the territorial extension.

***Indicator 6*:**

Published performance delivery standards for basic services – PPDS (yes/no)

"The indicator addresses the quality of institutions by measuring the existing mechanisms for effective delivery of various basic services" (UN-HABITAT, 2004a: 33). In particular, this indicator intends to assess the "presence or absence of a formal publication by the local government of performance standards for key services delivered by the local authority" (UN-HABITAT, 2003c: 9). This is an indicator concerning the efficiency of the local government in delivering services, its accountability to the targets it has set for itself, confidence of local population in the local government and equitable distribution of services by the local government (ibid.).

Indicator adapted to the case study of water supplies

Published performance delivery standards for water supply:

- *Is there currently a formal publication of performance standards for water supplies delivered by the local authority?*
- *At what institutional level does the publication of performance standards takes place? (municipality/district/state/province)*
- *What is the medium of publication for the standards? (e.g. Newspaper, radio, Internet, notice board, etc.)*

PAZ CASTILLO AND CRISTÓBAL ROJAS MUNICIPALITIES:

>>> data: no – result: 0

The municipalities of Paz Castillo and Cristóbal Rojas did not have a performance delivery standard for water supplies because – according to the interviewed public officials – the service was managed and organised by Hidrocapital.

HIDROCAPITAL:

On the contrary, Hidrocapital had a standard related to the "quality" of water and a specific office with a person in charge of its functioning (in Caracas and in Yare); this standard for the quality of water was established at the provincial level given that Hidrocapital includes the Federal District, the Vargas State and the Miranda State. This is due to the fact that Hidrocapital was still responsible for the distribution and management of water supply.

Indicator 7

Customer satisfaction survey – CCS (yes/no)

"The indicator addresses mechanisms in place that integrates civil concerns in improving service delivery, fostering effective governance.

... this indicator addresses the inclusion of citizens perception coherent to the effort to dilute local government bias in selecting indicators within the Effectiveness principle" (UN-HABITAT, 2004a: 33). In particular, it assesses the "existence and frequency of a survey on customers' satisfaction with the local authority's services. This indicates the willingness of the local administration to receive critical feedback from the local population and also the willingness to modify existing systems for working with the local population, which is likely to enable them to be more effective" (UN-HABITAT, 2003c: 10).

Indicator adapted to the case study of water supplies:

Customer satisfaction survey upon the water services:

- *Has a survey concerning customer satisfaction with regards to local government water service provisions been undertaken in the city?*
- *At what institutional level does the publication of performance standards takes place? (municipality/district/state/province)*
- *If yes, what has been the frequency of such a survey?*

PAZ CASTILLO AND CRISTÓBAL ROJAS MUNICIPALITIES:

>>> data: no – result: 0

The municipalities of Paz Castillo and Cristóbal Rojas did not have a customer satisfaction survey for water supplies because – according to the interviewed public officials – the service was still managed and organised by Hidrocapital.

HIDROCAPITAL:

Likewise, Hidrocapital did not organise a customer satisfaction survey for the water service. Nevertheless, Hidrocapital officials stated that consumers could always complain at the hydrological enterprise's offices.

Indicator 8

Vision statement[3] – VSE (yes/no)

"The indicator addresses the mechanisms in place for effective articulation of a city's goal" (UN-HABITAT, 2004a: 33). It represents

[3] "The city may have a vision statement developed by various stakeholders of the city. Like community associations, corporate sector, academicians, NGOs, etc., however statements by the local government is of utmost importance as it decides and implements most of the policies for its people" (UN-HABITAT, 2003c: 11).

"the measure of local authorities' commitment in articulating a vision for the city's progress. Does the local authority articulate a vision for the city's future through a participatory process?" (UN-HABITAT, 2003c: 11). The "existence of a vision statement implies a first step commitment of the local authority towards welfare of its people and the aspiration of its people. ... A vision statement articulated in a participatory manner increases its accountability and ownership" (ibid.).

PAZ CASTILLO MUNICIPALITY:

>>> data: no – result: 0

According to the interview with the Financial Director of the Paz Castillo municipality, the municipality had not yet developed a vision for the city's progress. Nevertheless, Paz Castillo was part of the project *Alcaldìa Digital*, endorsed by the Ministry of Science and Technology with the objective of "building the technological and managerial conditions that can allow municipalities to effectively process knowledge and information, in order to optimize local management quality and promote interactions with citizens" (http://www.alcaldias.gov.ve/portal-alcaldias/portal.html). In particular, this project requests municipalities to provide a vision and a mission concerning their future developments: in the case of Paz Castillo, the vision was not yet available on the website (last check: 27th of January 2007). In the *Alcaldìa Digital* website it was not specified whether this vision needs to be drafted through a participatory process.

CRISTÓBAL ROJAS MUNICIPALITY:

>>> data: 1 / 0 – result: 0.50

According to the interview with the Director of "Urban and Rural Planning", the municipality did not have a vision statement. Nevertheless, in the Alcaldia Digital website, the municipality of Cristóbal Rojas displayed a mission and a vision elaborated by its mayor[4]. However this

[4] *El trabajo gubernamental del gobierno que presido, se fundamenta en destinar todos los esfuerzos para ensanchar la base de la Participación Ciudadana, como mandato estelar de la Constitución Bolivariana. Obviamente, implementar una plataforma tributaria que permita sincerar la estructura de contribuyentes, la mayoría de ellos morosos con la municipalidad porque nunca se les cobró esas obligaciones, y desde luego, consolidar la distribución presupuestaria en programas sociales a cargo de la Dirección de Desarrollo Social y de obras públicas, bajo la responsabilidad de la Dirección de Ordenación Urbanística, son imperativos de esa visión. El actual gobierno de Charallave que dirijo como alcaldesa, está enfilado hacia objetivos de construcción de una ciudad moderna, no sólo porque el ejecutivo está a mí cargo, que como apegada a los principios que me invisten como profesional Urbanista, son razones fundamentales para impulsar esa visión, tanto en los social, como en la redistribución urbanística basada en el impacto de la obra ferroviaria, la cual desde ya está incidiendo como en ningún otro municipio del país, en el epicentro del*

vision statement presented in the website seems to be more of a personal commitment by the mayor and not drafted through a participatory process. This conclusion was also supported by the fact that the Director of "Urban and Rural Planning" did not know of the existence of the statement.

>>> **vision statement**: yes = 1

>>> **vision statement draft through a participatory process**: no = 0

Indicator adapted to the case study of water supplies

Existence of a vision statement related to the water sector:

- *Is there a vision statement developed for the urban water sector by the local government?*
- *If yes, has the vision statement been drafted through a participatory process (involving local government, civil society and the private sector)?*

HIDROCAPITAL:

The LOPSAPS can be considered as a sort of vision statement that articulates the goal of improved access to water and sanitation services. Even if it is not drafted in a participatory manner, the principle of participation is at the basis of this law. During a face-to-face interview (22-02-06), the Community Coordinator of the Losada Ocumarito System argued that the new Hidrocapital vision rests on the distribution and access to drinking water to all people, based on the assumption that human beings have the right to water. The General Manager of the Losada Ocumarito system stated that Hidrocapital has the objective (year's mission) to give water 365 days per year, 24 hours per day and to all people in its territorial jurisdiction.

Possible new indicator

Specific section/department charged with the organisation and management of water and sanitation services in the two selected municipalities (yes/no)

>>> data: no – result: 0

---

desarrollo nacional, tal como se concibe en la visión del gobierno nacional. La tarea es ardua, pero segura y sostenida, porque estoy convencida y compenetrada con el Plan de Desarrollo Nacional. Soy consecuente con la planificación y ha sido el norte de mí gestión de gobierno, y con mayor razón cuando la realidad nos presenta como un municipio con diversas variables altamente aprovechable para consolidar esta nueva encrucijada geopolítica del país: Charallave (http://cristobalrojas-miranda.gov.ve/portal-alcaldias/).

This indicator addresses the efficiency of local municipalities in organising and managing the water service provisions in their areas. In particular, it assesses the current institutional facility and human resources for the water services, given that with the LOPSAPS, the decentralisation of responsibilities for the water services to municipalities is envisaged.

Up to the time of investigation, the decentralisation process in the two studied communities had not started and no specific department, thus, had been established (see chapter 5).

<u>PAZ CASTILLO MUNICIPALITY</u>: effectiveness indicators

No.	Indicator	Data (X)	Formula	Result	Weight	Total
1	Local Government revenue per capita (LGR)	$ 86.08	LGR = (log X-log min)/ (log max-log min) min=2.3 max=1,340 (UN-HABITAT, Field test, 2003 US$)	0.57	0.25	0.14
2	Ratio of recurrent and capital budget (RRC)	0.16	RRC = (log X-log min)/ (log max-log min) min=0.09 max=8.37 (UN-HABITAT, Field test, 2003)	0.13	0.10	0.013
3	Local government revenue transfer (LGT)	86%	LGT = 1 (0-25%=1, 25-50%=0.75, 50-75%=0.50, 75-100%=0.25)	0.25	0.10	0.025
4	Ratio of mandated to actual tax collected (TC)	n. a.	TC = a/b	n. a.	0.10	n. a.
5	Predictability of transfers in local government budget (PoT)	Yes = 1	PoT = X	1.00	0.10	0.10
6	Published performance delivery standards (PPDS) for water supply	No = 0	PPDS = PPS × S/T	0.00	0.15	0.00
7	Consumer satisfaction survey (CSS) for water supply	No = 0	CSS = X	0.00	0.10	0.00
8	Vision statement effective (VSE)	No = 0	VSE = X	0.00	0.10	0.00
					Effectiveness sub-index	**0.309**

CRISTÓBAL ROJAS MUNICIPALITY: effectiveness indicators

No.	Indicator	Data (X)	Formula	Result	Weight	Total
1	Local Government revenue per capita (LGR)	123.9 $	LGR = (log X-log min)/(log max-log min) min=2.3 max=1,340 (UN-HABITAT, Field test, 2003 US$)	0.63	0.25	0.16
2	Ratio of recurrent and capital budget (RRC)	1.25	RRC = (log X-log min)/(log max-log min) min=0.09 max=8.37 (UN-HABITAT, Field test, 2003)	0.58	0.10	0.058
3	Local government revenue transfer (LGT)	44.4%	LGT = 1 (0-25%=1, 25-50%=0.75, 50-75%=0.50, 75-100%=0.25)	0.75	0.10	0.075
4	Ratio of mandated to actual tax collected (TC)	around 50%	TC = a/b	0.50	0.10	0.05
5	Predictability of transfers in local government budget (PoT)	Yes = 1	PoT = X	1.00	0.10	0.10
6	Published performance delivery standards (PPDS) for water supply	No = 0	PPDS = PPS × S/T	0.00	0.15	0.00
7	Consumer satisfaction survey (CSS)	No = 0	CSS = X	0.00	0.10	0.00
8	Vision statement effective (VSE)	Yes/No	VSE = X	0.50	0.10	0.05
					Effectiveness sub-index	**0.49**

EQUITY INDICATORS:

Indicator 9

Citizens' charter, right of access to basic services – CCS (yes/no)

"The indicator addresses institutional accountability towards citizens in providing equitable access to services" (UN-HABITAT, 2004a: 35). It is represented by the "existence or absence of a signed, published statement (charter) from the local authority which acknowledges citizens' right of access to basic services" (UN-HABITAT, 2003c: 12).

Indicator adapted to the case study of water supplies

Citizens' charter concerning the right of access to water supplies

- *Is there a signed, published statement (charter) from the local authority which acknowledges the citizens' right of access to water supplies?*
- *At what institutional level does the published statement acknowledge the citizens' right of access to water supplies?*
- *What is the medium of publication for the charter? (e.g. Newspaper, radio, Internet, notice board, etc.)*

PAZ CASTILLO AND CRISTÓBAL ROJAS MUNICIPALITIES:

>>> data: no – result: 0

In a certain way, the Constitution (1999) acknowledges the right of access to water at the national level: "*All waters are goods in the public domain belonging to the Nation and indispensable for life and development*" (art. 304). Nevertheless, it is difficult to consider this as a "charter" in regards to the right of access to water supplies. Interviews with Mercedes de Freitas (Transparencia Venezuela) and Maria Elena Rodriguez (PROVEA) confirmed that Venezuela had not yet been provided with a charter concerning the right of access to water supplies, at the time of investigation (excluding thus the 2008 *Ley de Agua*).

Indicators 10 and 11

Percentage of women councillors in local authorities – WC / WK (%)

"The indicator addresses gender equity in representation of women involved in local government decision making" (UN-HABITAT, 2004a: 35). "Equal gender representation in local government is regarded as fundamental to promoting more equitable policies, plans and projects" (UN-HABITAT, 2003c: 13).

Two variables define this indicator:

- *Women councillors as a percentage of the total number of councillors in a local authority (in the last election) (with higher loading: 0.75)*
- *Women councillors in key positions (with lower loading: 0.25)*

PAZ CASTILLO AND CRISTÓBAL ROJAS MUNICIPALITIES:

>>> data (WC): 28.58% – result: 0.57

>>> data (WK): not available

Indicator 10. In the Paz Castillo municipality, according to the municipal elections in 2005 (last municipal election before 2006), there were 7 councillors for each municipality and, in both case, 2 of them are women. (Source: *La Voz*, 09/08/2005)

>>> **WC: 28.58%**

Indicator 11. no data is available for the number of women councillors employed in key positions

>>> **WK: not available**

Indicator 12

(Indicator already adapted to the case of water supplies)

Pro-poor pricing policy for water:

"This indicator signifies commitment and measures undertaken by the local government for equitable distribution of basic services (water as a key service) to its entire people commensurate with there economical conditions, especially the poorest sections" (UN-HABITAT, 2003c: 15). As it is underlined in the Second Field Test Report, "it is difficult to arrive at one most convincing indicator to measure equity towards pricing policy. Thus, a combination of information is required to measure equitable access to water, subsidised water, water consumption slabs and comparison of residential water price to commercial or industrial sector" (UN-HABITAT, 2004a: 36). Thus, with regard to this indicator, three more different indicators are recommended:

PAZ CASTILLO AND CRISTÓBAL ROJAS MUNICIPALITIES:

Indicator 12A

Proportion of households with direct access to water (piped connections) in the two selected municipalities – HH wat (%)

>>> 74.82% – result: 0.75

The proportion of the networked water in the two selected municipalities was: Cristóbal Rojas: 92.21%; Paz Castillo: 74.82% (see the following table based on the 2001 Census). Nevertheless, it is worth pointing out that according to Lacabana *et al.* (2004) the figures, first, are overestimated and, second, do not take into account the problems related to the quality and frequency of the service. Furthermore, according to UN-HABITAT (2004b: 15, indicator No. 6: connection to services), it could be useful to separate the specific results for households living within the informal settlements of these two municipalities (that is, Paz Castillo and Paso Real 2000) from the entire municipal area, in order to have a more appropriate picture of water access in low-income communities.

Caracas Metropolitan Region			**Municipalities**	**Household**	**Water**		
					Networked Water	**Water – Trucks**	**Other**
CMR	PUI	CMA		680,196	97.38	0.47	2.15
CMR	PUI	Guarenas – Guatire		82,110	92.38	2.99	4.63
CMR	PUI	TUY	***Cristóbal Rojas***	*18,318*	*92.21*	*3.35*	*4.44*
CMR	PUI	TUY	Independencia	27,947	95.19	2.06	2.75
CMR	PUI	TUY	Lander	24,486	88.97	2.41	8.62
CMR	PUI	TUY	***Paz Castillo***	*19,375*	*74.82*	*18.38*	*6.80*
CMR	PUI	TUY	Simón Bolívar	7,171	92.23	2.69	5.08
CMR	PUI	TUY	Urdaneta	24,405	92.88	3.21	3.91
CMR	PUI	TUY	***Total***	***121,702***	***89.61***	***5.19***	**5.20**
CMR	PUI	Altos Mirandinos		69,479	90.71	0.62	8.67
CMR	PUI	Vargas		71,260	87.35	3.13	9.52
CMR	PUI	***Total RMC***		***1,024,747***	***94.91***	***1.43***	**3.67**

Source: adapted from Lacabana (2003), based on the national Census of 2001.

Indicator 12B

Is there a pro-poor pricing policy for water? – PPC (yes/no)

>>> data: yes – result: 1

The LOPSAPS can also be considered as a water pricing policy, given that it includes tariff structures for the service. In this law it is stated that the tariff model will be defined by the *Superintendencia Nacional de los Servicios de Agua Potable y Saneamiento* and approved by municipal authorities (art. 83). At the moment of investigation, the tariff structure was ruled by the *Régimen Tarifario para la Prestación de los Servicios de Acueductos y de Recolección, Tratamiento y Disposición de Aguas Residuales* (see below). The LOPSAPS states that the tariff has to include all the administrative, operative and maintenance costs together with the costs concerning investments in rehabilitation, expansion and capital's remuneration (art. 85). In addition, the tariff is supposed to be composed by a fix and a variable charge – where the latter reflects the effective quantity of water consumption (art. 86).

In addition, the new law for water and sanitation envisages different kinds of subsidies in order to "finance minimum consumption necessary for a household" and "focus on the households with scarce resources" (art. 95). These subsidies can be conceded at the "demand" – which is a financial aid to the subscriber –, at the "offer" – which is a financial support to investments – or "cross-cutting subsidies" envisaging a higher tariff to partly cover the payment of low-income households (art. 96-101). They can be applied by the national executive power, by the states, by the municipalities and by the metropolitan districts. The LOPSAPS explicitly aims at allowing households unable to pay the entire cost of water to benefit as equally as all citizens in need of the water service (Baez, 2002).

Indicator 12C

Is the metered price of water per litre lower for households below the poverty line? – WP (yes/no)

>>> data: yes – result: 1

The LOPSAPS states that the *Supeintendencia Nacional de los Servicios de Agua Potable y Saneamiento* has to "establish the subscribers' categories associated with the tariff structures on the basis of their economic activities and socio-economic activities of residential subscribers" (art. 87). At the time of the fieldwork, Hidrocapital had different levels of tariffs for water consumption according to the building's use destination and income population's levels. The Gaceta Oficial No. 35161 of the 1st March 1993 *Régimen Tarifario para la Prestación de los Servicios de Acueductos y de Recolección, Tratamiento y Disposición de Aguas Residuales* (further updated by the new tariff regime in the Gaceta Oficial No. 39788, published on the 28th of November 2011) classifies four "modes of use": 1. residential; 2. social residential 3. commercial; 4. industrial. In particular, the second one includes buildings "located in geographic areas where a wide part of low-income population lives" and that can be recognised as eligible for the application of the social tariff according to the income levels (up to 40 m^3/month). In addition, each mode of use comprises 6 different tariffs given by the operation costs connected with the water and sanitation infrastructure and the hydrological capacity. In addition, the subsidies' regime allows low-income households to pay less for water.

Indicator 13

Incentives for informal business – IA (yes/no)

"The indicator addresses the efforts of government to provide opportunities for informal business" (UN-HABITAT, 2004a: 36). This indicator, addressing an informal activity, is supposed to be difficult to quantify in absolute terms; thus, three variables help assessing the local government effort to support the informal sector:

- *street vending not allowed*
- *street vending with restrictions*
- *public fairs, municipal markets*

PAZ CASTILLO AND CRISTÓBAL ROJAS MUNICIPALITIES:
>>> data: yes – result: 1

Indicator adapted to the case study of water supplies

Incentives for the development of informal arrangements (as the mesa tecnica de agua) to access water supplies:

- *Are informal provisions allowed by the local government?*

In the case of illegal connections, for instance, in the two selected communities and in the case of water trucks.

- *Are there restrictions in the informal access to water?*

Perhaps the new LOPSAPS can be seen as a restriction for informal access to water, in the sense that citizens are asked to organise themselves in order to improve their access to water.

- *Is there an explicit effort to support informal access to water?*

Yes, because, through the MTA and the LOPSAPS, illegal connections to the main system need to be better organised and further developed rather than removed, in order to improve access to water in low-income urban communities. It could be interesting to see how the LOPSAPS continues this process in which informal settlements are not removed, but rather their access to services is improved.

PAZ CASTILLO MUNICIPALITY: equity indicators

No.	Indicator	Data (X)	Formula	Result	Weight	Total
9	Citizens charter for water services (CCS)	No = 0	CC = X	0.00	0.20	0.00
10	Percentage of women councillors (WC)	28.58%	WC = X × 2/100	0.57	0.20	0.114
11	Percentage of women in key positions (WK)	n.a.	WK = X × 2/100	n.a.	0.10	n.a.
12A	Percentage of households with water connection (HH wat)	74.82%	HH wat = 74.82/100	0.75	0.15	0.112
12B	Existence of pro-poor policy (PPC)	Yes = 1	PPC = X	1.00	0.10	0.10
12C	Is water price cheaper for poor settlements? (WP)	Yes = 1	WP = X	1.00	0.10	0.10
13	Incentives for informal arrangements to improve water access (IA)		IA = 1 (any one of a, b or c)	1.00	0.15	0.15
	a. Informal provisions allowed	Yes = 1				
	b. Informal provisions with restrictions	No = 1				
	c. support to informal provisions	Yes = 1				
					Equity sub-index	**0.63**

CRISTÓBAL ROJAS MUNICIPALITY: equity indicators

No.	Indicator	Data (X)	Formula	Result	Weight	Total
9	Citizens' charter for water services (CCS)	No = 0	CC = X	0.00	0.20	0.00
10	Percentage of women councillors (WC)	28.58%	WC = X × 2/100	0.57	0.20	0.114
11	Percentage of women in key positions (WK)	n.a.	WK = X × 2/100	n.a.	0.10	n.a.
12A	Percentage of households with water connection (HH wat)	92.21%	HH wat = 92.21/100	0.92	0.15	0.138
12B	Existence of pro-poor policy (PPC)	Yes = 1	PPC = X	1.00	0.10	0.10
12C	Is water price cheaper for poor settlements? (WP)	Yes = 1	WP = X	1.00	0.10	0.10
13	Incentives for informal arrangements to improve water access (IA)		IA = 1 (any one of a, b or c)	1.00	0.15	0.15
	a. Informal provisions allowed	Yes = 1				
	b. Informal provisions with restrictions	Yes = 1				
	c. support to informal provisions	Yes = 1				
					Equity sub-index	**0.67**

PARTICIPATION INDICATORS:

Indicator 14 *(representative democracy)*

Elected council[5] – EC (yes/no)

"The indicator measures whether the local governing council is elected through a democratic process or not. ... An elected council is more likely to make decisions that are more representative of the wishes of the local population" (UN-HABITAT, 2003c: 17).

- *Are the councillors locally elected?*
- *Are the councillors both elected and appointed?*

PAZ CASTILLO AND CRISTÓBAL ROJAS MUNICIPALITIES:

>>> data: yes – result: 1

Councillors are democratically elected in all of the Venezuelan municipalities, as per the *Ley Organica del Regime Municipal* 1989 and the 2005 *Ley Organica del Poder Publico Municipal* (LOPPM – Municipal Public Power Act).

Indicator 15 *(representative democracy)*

Election of Mayor[6] – LEM (yes/no)

"The indicator addresses representative decision-making ... However, to avoid penalisation of systems where the Mayor is not directly elected, it was advisable to have intermediate scores for this indicator" (UN-HABITAT, 2004a: 38):

Method of Mayor's Selection	score
Directly elected	*1.00*
Elected amongst councillors	*0.75*
Appointed	*0.50*

PAZ CASTILLO AND CRISTÓBAL ROJAS MUNICIPALITIES:

>>> data: yes, directly elected – result: 1

5 "This information should be verified through urban experts and urban policy makers involved in the city planning. They may be from the planning department of the city or the National Ministry responsible for urban development. Any important change which has occurred during the last five years should be mentioned and explained" (UN-HABITAT, 2004b: 51).

6 *Idem.*

Mayors are democratically and directly elected on the base of the *Ley Organica del Regime Municipal* (1989) and the 2005 *Ley Organica del Poder Publico Municipal* (LOPPM – Municipal Public Power Act). Before these laws, the mayor (president of municipal council) was elected amongst councillors or appointed by the governor.

Indicator 16 *(representative democracy)*

Percentage voter turnout – VT (%)

It is defined as the "percentage of adult (male and female) population (having reached voting age) who voted in the last municipal elections. ... The indicator measures the degree of interest and involvement of the public in local government" (UN-HABITAT, 2004b: 52).

The last municipal elections for the *consejo communal*, before 2006, took place on the 7th of August 2005. The website of the Consejo Nacional Electoral (www.cne.gov.ve) provides the data concerning the percentage of abstention in each Venezuelan state. The percentage of voter turnout is not available for the single municipalities, but just for each Venezuelan state. In the case of the Miranda state – where the two municipalities are located – the percentage of abstention is 75.89%. Consequently, the percentage of voter turnout can be reckoned to be at 24.11%. Nevertheless, *La Voz* (09/08/2005) provides a specific figure for the percentage of voter abstention in Paz Castillo (80.49%), identifying it as the higher percentage of abstention in the Tuy Valleys. This implies a voter turnout of 19.51% in Paz Castillo municipality. Eventually, the last figure is chosen as the data of this indicator. The last Mayor and Governor elections, before 2006, took place on October 31, 2004. The website of the Consejo Nacional Electoral provides the data concerning the percentage of abstention in each Venezuelan municipality. Final data concerning voter turnout is considered as the average between the mayoral elections in 2004 and municipal elections in 2005 for each municipality.

PAZ CASTILLO MUNICIPALITY:

>>> elections 2004 (mayor) data: 48.57% – result: 0.49

>>> elections 2005 (municipal elections) data: 19.51% – result: 0.20

>>> average (2004 and 2005) data: 34.04% – result: 0.34

CRISTÓBAL ROJAS MUNICIPALITY:

>>> elections 2004 (mayor) data: 51.2% – result: 0.51

>>> elections 2005 (municipal elections) data: 24.11% – result: 0.24

>>> average (2004 and 2005) data: 37.66% – result: 0.38

Indicator 17 *(participative democracy)*

Public forum[7] – PC (yes/no)

"The public forum could include people′ council, city consultation, neighbourhood advisory committees, town hall meetings, etc. ... The existence of a people's forum indicates the availability of informal of formal mechanisms for the public to express their views and share their issues with peers" (UN-HABITAT, 2003c: 20).

Public forum in the two selected municipalities:

- *Is there any public forum/council for the citizens to express their views?*

PAZ CASTILLO AND CRISTÓBAL ROJAS MUNICIPALITIES:

>>> data: yes – result: 1

It is possible to observe that, at the local level, there are many public committees and associations through which the community organises itself, such as the *asociacion civil, asociacion de vecinos, comité de tierra urbana* (CTU), *mesa tecnica de agua* (MTA), *mesa tecnica de luz* (MTL), *comité de controleria*, etc. Nevertheless, the "public forum" legally recognised by the law can be identified in the *Consejos Locales de Planificacion Publica* (Local Public Planning Council – LPPC) which involved representatives of the above associations (Constitution, 1999; art. 182). They have the objective to include citizens in the decision making process and involve the community's representatives belonging to the different sectors of the organised civil society: education, health, culture, sport, production and trade, transport, ecology, services and natural resources (art. 4, *Ley de los Consejos Locales de Planificacion Publica*, Gaceta Oficial No. 37463, 12th of June 2002). The LPPCs should be constituted by the mayor, the councillors, the chairman of the parish forum and the people in charge of community or neighbour association (present in the same quantity plus one of the previous members). According to Lacabana *et al.* (2004: 25), "the new Constitution approved in 1999 incorporates a set of legal instruments and governmental and non-governmental arrangements designed to advance a process of democratic decentralisation and creates new channels of negotiation and consensus-building … A new local institution, the Local Public Planning Council (LPPC) is responsible for the local government's comprehensive planning and for ensuring co-ordination and popular participation in the drafting and monitoring of the Municipal Development Plan and the programs and actions taken in the Municipality".

More recently, other laws promoted other forms of citizens' councils, such as the communal councils (*Ley Organica de los Consejos Comunales*, 2007 and 2009) and the communes (*Ley de las Comunas*, 2010).

7 *Idem.*

Indicator 18 *(participative democracy)*

Civic Associations per 10,000 population[8] – CA (No.)

>>> not available (>>> see for instance www.venescopio.org.ve and www.cisor.org.ve)

This indicator is defined as the "number of voluntary non-profit organisations, including NGOs, political sporting or social organisations, registered or with premises in the city, per 10,000 population. ... Civic organisations can help ensure the accountability of local government ... may facilitate the formation of partnerships for the delivery and/ or maintenance of services" UN-HABITAT, 2004b: 53). This indicator "measures the vibrancy of the associational life in a city ..." (UN-HABITAT, 2003c: 20).

Indicator adapted to the case study of water supplies

alternative 3 *(new possible indicator)*:

Mesas tecnicas de agua in the two selected municipalities: this indicator can provide a measure of citizen participation through the associational life, in order to improve their access to water supplies. If the figures are available, it is possible to compare the data of associational life in the two selected municipalities with the data in the other municipalities of the Metropolitan Region of Caracas.

Nevertheless, in the Valleys of the Tuy Medio there were not official data, According to the Community Coordinator of the Losada Ocumarito System (interview 22-02-06), about 52 MTAs were active in 2006.

PAZ CASTILLO MUNICIPALITY: participation indicators

No.	Indicator	Data (X)	Formula	Result	Weight	Total
14	Elected council (EC)	Yes = 1	EC = X	1.00	0.15	0.15
15	Locally elected mayor (LEM)	Yes = 1	LEM = X	1.00	0.15	0.15
16	Voter turnout (VT)	34.04%	VT = x/100	0.34	0.30	0.10
17	Public forum (PC)	Yes = 1	PF = X	1.00	0.15	0.15
18	Civic associations per 10,000 pop (CA)	X = n. a.	CA = (Log X – Log 0.49)/(Log 72.79 – Log 0.49) min= 0.49; max= 72.79 (UN-HABITAT, Field test, 2003)	n. a.	0.25	n. a.
					Participation sub-index	**0.73**

8 "This information may be obtained through umbrella associations of NGOs and key NGOs aware of the number of associations in each particular area. Different Ministries may be consulted which may maintain registries of NGOs" (UN-HABITAT, 2004b).

<u>CRISTÓBAL ROJAS MUNICIPALITY</u>: participation indicators

No.	Indicator	Data (X)	Formula	Result	Weight	Total
14	Elected council (EC)	Yes = 1	EC = X	1.00	0.15	0.15
15	Locally elected mayor (LEM)	Yes = 1	LEM = X	1.00	0.15	0.15
16	Voter turnout (VT)	37.66%	VT = x/100	0.38	0.30	0.11
17	Public forum (PC)	Yes = 1	PF = X	1.00	0.15	0.15
18	Civic associations per 10,000 pop (CA)	X = n. a.	CA = (Log X – Log 0.49)/(Log 72.79 – Log 0.49) min= 0.49; max= 72.79 (UN-HABITAT, Field test, 2003)	n. a.	0.25	n. a.
					Participation sub-index	**0.75**

ACCOUNTABILITY INDICATORS:

Indicator 19 *(transparency)*

Formal publication of contracts, tenders, budget and accounts[9] – FP (yes/no)

"The indicator measures the formal procedures that foster transparency in the local government operations" (Un-habitat, 2004a: 39). "The open flow of information is vital for good urban governance. The formal publication of information regarding contracts and budgets indicates the willingness of the local authority to be transparent in its activities and accountable for its decisions. It would also indicate greater confidence of people in the local government and integrity of auditing and monitoring" (UN-HABITAT, 2003c: 22).

- *Is there a formal publication of:*
  - *contracts and tenders?*
  - *budgets and accounts?*
- *What is the medium of publication for these documents? (e.g. Newspaper, radio, Internet, notice board, etc.)*

According to the *Ley contra la Corrupcion* (2003), all the information about contracts, tenders, budgets and accounts concerning public patrimony has a public character (art. 8). As a result, the municipality "shall inform citizens about the use of goods and the expenditures related to the public patrimony resources. For this reason, every three months, [the municipality] shall publish and let accessible to all citizens in specific offices (*oficinas de atencion al publico o de atencion ciudadana*) a detailed report, easy to handle and understand, over the administrated patrimony and the relative description and justification of its use. ..." (art. 9). In general, given that the information about the formal publication of this report was not available, for the calculation of this indicator we consider that, if the *Oficina de Atencion Ciudadana* (belonging to the *Contraloria General de la Republica*) or the *Contraloria Municipal* had already been set up in the municipality, the access to this information is easier.

[9] "Information should be obtained from the local authorities' officials or the Ministry responsible for Local Governments and checked with representative of NGOs dealing with governance issues. There should be institutionalised mechanisms which ensure that contracts and tenders for municipal services are always published in order to ensure an adequate competitive transparent and open process in contracting" (UN-HABITAT, 2004b: 54).

PAZ CASTILLO MUNICIPALITY:

>>> data: no/no – result: 0

Given that the *Oficina de Atencion Ciudadana* (belonging to the *Contraloria General de la Republica*) had not yet been set up in the Paz Castillo municipality, contracts, tenders, budgets and accounts were not formally published and, thus, not easily consulted. Nevertheless, according to the interview with the Financial Director of the Paz Castillo municipality, a copy of tenders and budgets was freely available at the municipal offices (*Camara Municipal*).

CRISTÓBAL ROJAS MUNICIPALITY:

>>> data: not available

One copy of contracts, tenders, budgets and accounts was deposited at the *Oficina de Atencion Ciudadana y Contraloría Social de la Contraloría Municipal del Municipio Cristóbal Rojas* (belonging to the *Contraloria General de la Republica*). Information about the formal publication of contracts, tenders, budgets and accounts through a specific medium – such as newspapers, radio, notice boards – was not available, but these documents were easily accessible at the citizen office (information collected through the Director of "Urban and Rural Planning").

***Indicator 20** (responsiveness)*

Control by higher level of government[10] – CG (yes/no)

"The indicator measures responsiveness in local governance" (UN-HABITAT, 2004a: 40). It "measures the control of higher levels of government (National, State/Provincial) for closing the local government and removing councillors from office. Indicates the direction of accountability. If central government is able to close local authorities or remove councillors at its discretion, councillors are likely to be more accountable upwards rather than downwards to their citizens" (UN-HABITAT, 2003c: 22-23).

This indicator can also be considered as a measure of the level of the decentralisation process (UN-HABITAT, 2004b: 50, check-list 7: decentralisation).

[10] Information should be obtained from the local authorities officials. Local government comprises all local-level governments, whether they be (UN-HABITAT, 2004b: 50): "legally, fiscally, and politically autonomous jurisdiction; centrally influenced but legally distinct local governments (municipality with centrally appointed mayors); some forms of quasi-government organisations which exists to deliver service locally".

PAZ CASTILLO AND CRISTÓBAL ROJAS MUNICIPALITIES:

- *Can higher levels of government (National, State/Provincial)...:*

>>> **data: no/no – result: 1**

• *... close the local government?*

No, municipalities are autonomous at an administrative and managerial level with respect to higher levels of government, as per the Venezuelan Constitution (1999) art. 168[11] and the *Ley Organica del Poder Publico Municipal* – LOPPM, 2005.

• *... remove councillors from office?*

No, municipalities are autonomous at an administrative and managerial level with respect to higher levels of government.

- *Can the local government, without permission from higher governments*:

>>> **data: yes/no/n.a./yes – result: 0.67**

• *... set local tax levels?*

Yes, according to the *Ley Organica del Poder Pùblico Municipal* (2005), art. 162.

• *Set user charges for services (water)?*

No, because in the case of drinking water the service is organised, distributed and managed by Hidrocapital and the charges/prices decided and collected by Hidrocapital. In the same way, according to the new LOPSAPS (2001) a final control by the National Superintendency of Drinking Water and Sanitation Services is requested. Nevertheless, the *Ley Organica del Poder Pùblico Municipal* (2005) states that municipalities can set charges for the services they provide.

• *... borrow funds?*

n.a. According to the interviews in the administrative sectors in both the municipalities, it seems that the two selected municipalities have never faced the problem of borrowing funds from external agencies.

---

11 *Artículo 168. Los Municipios constituyen la unidad política primaria de la organización nacional, gozan de personalidad jurídica y autonomía dentro de los límites de esta Constitución y de la ley. La autonomía municipal comprende: 1. La elección de sus autoridades; 2. La gestión de las materias de su competencia; 3. La creación, recaudación e inversión de sus ingresos. Las actuaciones del Municipio en el ámbito de sus competencias se cumplirán incorporando la participación ciudadana al proceso de definición y ejecución de la gestión pública y al control y evaluación de sus resultados, en forma efectiva, suficiente y oportuna, conforme a la ley. Los actos de los Municipios no podrán ser impugnados sino ante los tribunales competentes, de conformidad con esta Constitución y con la ley (Constitución de la República Bolivariana de Venezuela, 1999).*

- *... bhoose contractors for projects?*

Yes, but respecting the national legislation, according to the *Ley Organica del Poder Pùblico Municipal* (2005), art. 73 and 74.

Indicator 21 *(integrity)*

Codes of conduct – CoC (yes/no)

"The indicator addresses governments' commitment towards integrity of officials. Codes of conduct per se are often incorporated into the local government regulations at the state/provincial level and not always as a stand-alone formal document" (UN-HABITAT, 2004a: 40).

- *Is there a signed, published statement of the standards of conduct that citizens are entitled to from their elected officials and local government staff?*
- *At what institutional level are these codes of conduct prescribed? (Municipality, State/Province)*
- *If the codes of conduct are prescribed at the District/State/Province level are they applied to local councillors?*

PAZ CASTILLO AND CRISTÓBAL ROJAS MUNICIPALITIES:

>>> data: yes – result: 1

The administrative sector's Director of the Paz Castillo municipality and the Director of "Urban and Rural Planning" of the Cristobal Rojas municipality affirmed that standards of conduct for elected officials and local government staff were not collected in a single document but dispersed throughout different national laws and several local ordinances. In particular, the laws concerning the codes of conduct are: *Ley del estatuto de la function publica* (2002), *Ley Organica de la Contraloria General de la Republica y del Sistema Nacional de Control Fiscal* (2001), *Ley Organica del Poder Pùblico Municipal* (2005), *Ley contra la corrupcion* (2003, art. 6): *'En la administración de los bienes y recursos públicos, los funcionarios y empleados públicos se regirán por los principios de honestidad, transparencia, participación, eficiencia, eficacia, legalidad, rendición de cuentas y responsabilidad'.*

Indicator 22 *(integrity)*

Facility for citizen complaints[12] – FRC (yes/no)

"The existence of a facility established within the local authority to respond to complaints and a local facility to receive complaints

[12] "Information should be obtained from the local authorities' officials or the ministry responsible for local governments and checked with representative of NGOs dealing with governance issues" (UN-HABITAT, 2004b: 54).

and information on corruption. ... Indicators measure government's responsiveness toward integrity of officials and show the willingness of the government to be responsive for the welfare of its citizens" (UN-HABITAT, 2003c: 25).

- *Is there an appointed official to receive and respond to complaints against public authorities?*
- *Is there an exclusive facility to receive complaints on corruption?*

According to the new institutionalised Citizen Power (*Poder Ciudadano*) and the Republic General Comptrollership Unit (*Contraloria General de la Republica*), a specific office (*Oficina de Atencion Ciudadana*) is charged with informing citizens about public goods' use and public resources' expenditures, providing a detailed report about the public patrimony administration available to all people, and collecting and processing citizens' complaints and suggestions[13]. Citizens can go to these offices to make complaints.

PAZ CASTILLO MUNICIPALITY:

>>> data: no – result: 0

The Financial Director affirmed that in the Paz Castillo municipality there was no specific service for receiving citizens' complaints.

CRISTÓBAL ROJAS MUNICIPALITY:

>>> data: yes – result: 1

According to the interview with the Director of "Urban and Rural Planning", Cristóbal Rojas municipality was provided with *Oficina de Atencion Ciudadana* to receive complaints from citizens and to manage information. This office was also listed in the website of the Republic General Accounting Unit (www.cgr.gov.ve).

HIDROCAPITAL:

Hidrocapital has organised many commercial offices (*oficinas comerciales*) to collect drinking water payments and complaints regarding the services. According to our interviews with Hidrocapital officials, it

[13] *'atender y orientar a quienes asistan a la Contraloría General de la República para la resolución de un asunto determinado y de recibir, valorar y admitir, cuando sea procedente, las denuncias, reclamos, sugerencias y peticiones que formulen los ciudadanos cuya atención sea competencia de la Contraloría y, en el caso de no resultar manifiestamente improcedentes, remitirlas a los órganos correspondientes. Igualmente promoverá la participación y el control ciudadano, así como la educación como proceso creador de la ciudadanía' ... 'la obligación de todos los entes y organismos del poder público de crear las oficinas de atención al público o de atención ciudadana, donde se informará sobre la utilización de los bienes y el gasto de los recursos que integran el patrimonio público y que se debería poner a la disposición de cualquier persona un informe detallado de fácil manejo y comprensión, sobre el patrimonio que administran con la descripción y justificación de su utilización (www.cgr.gov.ve).*

seemed that citizens' complaints could be addressed to Hidrocapital only if the community or the person making the complaint was already paying for the water service. In addition, Hidrocapital receives complaints though the hotline 0800-potable, which are later archived with the help of an electronic service.

***Indicator 23** (corruption: disincentives and protection)*

Anti-corruption commission[14] – ACC (yes/no)

The indicator assesses the "existence of a local agency to investigate and report case of corruption. The anti-corruption commission shows the willingness of the government to scrutinize its own officials and protect its integrity by removing corrupt officials" (UN-HABITAT, 2003c: 26).

According to art. 41 of the Anti-corruption Law (2003, *Ley contra la Corrupcion*), the Republic General Comptrollership Unit (*Contraloria General de la Republica*) can investigate public officials and report cases of corruption.

PAZ CASTILLO MUNICIPALITY:

>>> data: no – result: 0

The Financial Director affirmed that in the Paz Castillo municipality the anti-corruption commission should be organised within the system of the government accounting inspectorate, as stated by the *Ley Organica de la Contraloria General de la Republica y del Sistema Ncional de Control Fiscal* (2001) and the *Ley contra la Corruption* (2003). Nevertheless he affirmed that, in the Paz Castillo municipality, there was no Municipality Comptrollership Office (*Controleria Municipal*) yet.

CRISTÓBAL ROJAS MUNICIPALITY:

>>> data: yes – result: 1

According to the interview with the Director of "Urban and Rural Planning", the agency to investigate and report cases of corruption was given by the Municipality Comptrollership Office (*Controleria Municipal*), which had already been established in Cristóbal Rojas.

***Indicator 24** (corruption: disincentives and protection):*

Disclosure of personal income assets[15] – PIA (yes/no)

"The indicator would represent the accountability of the to be decision-makers in the government and their genuine interest in the welfare of local people" (UN-HABITAT, 2003c: 27).

14 *Idem.*

15 *Idem.*

The Financial Director affirmed that in the Paz Castillo municipality and the Director of "Urban and Rural Planning" affirmed that in both municipalities, prior to entering and leaving an elected or nominated post, all public servants have to submit a public statement of income and assets. This is consistent with the art. 23 of the Anti-corruption Law (2003, *Ley contra la Corrupcion*).

PAZ CASTILLO AND CRISTÓBAL ROJAS MUNICIPALITIES:

>>> data: yes/no/no – result: 0.38

- *Prior to taking office, are locally elected officials required by law to publicly disclose their personal income and assets? (declaracion jurada de biene*s)
- **>>> data: yes – result: 1**
- *Prior to taking office, are locally elected officials required by law to publicly disclose their family income and assets?*

  >>> data: no – result: 0
- *Are local office bearers' income and assets regularly monitored?*

  >>> data: no – result: 0

***Indicator 25** (corruption: disincentives and protection)*

Regular independent audit[16] – RIA (yes/no)

"This indicator shows the accountability of the local government towards its taxpayers and transparency in providing resources for development projects" (UN-HABITAT, 2003c: 28).

- *Is there a regular independent audit of municipal accounts?*
- *Is the audit internal or external?*

On the basis of art. 42 and 57 of the Organic Republic General Comptrollership Unit and National Fiscal Control System (2001, *Ley Organica de la Contraloria General de la Republica y del Sistema Nacional de Control Fiscal*) and art. 100 of the Organic Public Power Act (2005, *Ley Organica del Poder Pùblico Municipal*), the Municipality Comptrollership Office (*Controleria Municipal*) shall organise an audit of municipal accounts every 5 years.

16 "Information should be obtained from the local authority's officials or the ministry responsible for local governments and checked with representative of the NGOs dealing with governance issues. Auditing of municipal account should be regular (annual or biannual). They should also be conducted by independent auditors" (UN-HABITAT, 2004b: 54).

PAZ CASTILLO MUNICIPALITY:

>>> data: no – result: 0

The Financial Director affirmed that in the Paz Castillo municipality, a Regular Independent Audit did not occur, because the *Contraloria Municipal* had not yet been instituted as stated by the current laws.

CRISTÓBAL ROJAS MUNICIPALITY:

>>> data: yes – result: 1

The Director of "Urban and Rural Planning" affirmed that the municipality carried a regular audit of municipal accounts, although it seems to be conducted internally rather than externally, according to art. 176 of the 1999 Constitution.

PAZ CASTILLO MUNICIPALITY: accountability indicators

No.	Indicator	Data (X)	Formula	Result	Weight	Total
19	Formal Publication (FP)		CTBA = Average (CT + BA)	0.00	0.20	0.00
	a. Formal publication: contracts and tenders (CT)	*No = 0*	*CT =X*	*0.00*		
	b. Formal publication: budget and accounts (BA)	*No = 0*	*BA = X*	*1.00*		
20A	Control by higher Govt. (CG)		CG = Average (CLG+RC)	1.00	0.07	0.07
	a. Control by higher Govt.: close local government (CLG)	*No = 1*	*CLG = X*	*1.00*		
	b. Control by higher Govt.: removal of councillors (RC)	*No = 1*	*RC = X*	*1.00*		
20B	Local government authorities (LGA)		LGA = Average (SLT+SYC+BF+CP)	0.67	0.08	0.05
	c. Local government: set local tax levels (SLT)	*Yes = 1*	*SLT = X*	*1.00*		
	d. Local government: set user charges for services (SUC)	*No = 0*	*SUC = X*	*0.00*		
	e. Local government: borrow funds (BF)	*n.a.*	*BF = X*	*n.a.*		
	f. Local government: choose contractors for projects (CP)	*Yes = 1*	*CP = X*	*1.00*		
21	Codes of conduct (CoC)	Yes = 1	CoC= X	1.00	0.10	0.10
22	Facilities to receive complaints (FRC)		FRC = Average (OA + EF)	0.00	0.10	0.00

No.	Indicator	Data (X)	Formula	Result	Weight	Total
	a. Official appointed to receive complaints on public authorities (OA)	*No = 0*				
	b. Exclusive facility to receive complaints on corruption (EF)	*No = 0*				
23	Anti-corruption commission (ACC)	No = 0	ACC = X	0.00	0.15	0.00
24	Personal Income and assets (PIA)		PIA = (0.75* Average PIA + FIA) + 0.25* IAM	0.38	0.15	0.06
	a. Disclosure of personal income and assets (PIA)	*Yes = 1*	*PIA = X*	*1.00*		
	b. Disclosure of family's income and assets (FIA)	*No = 0*	*FIA = X*	*0.00*		
	c. Income and assets regularly monitored (IAM)	*No = 0*	*IAM = X*	*0.00*		
25	Regular independent audit (RIA)	No = 0	RIA = X	0.00	0.15	0.00
					Accountability sub-index	**0.28**

<u>CRISTÓBAL ROJAS MUNICIPALITY</u>: accountability indicators

No.	Indicator	Data (X)	Formula	Result	Weight	Total
19	Formal Publication (FP)		CTBA = Average (CT + BA)	*n.a.*	0.20	*n.a.*
	a. Formal publication: contracts and tenders (CT)	*n.a.*	*CT =X*	*n.a.*		
	b. Formal publication: budget and accounts (BA)	*n.a.*	*BA = X*	*n.a.*		
20A	Control by higher Govt. (CG)		CG = Average (CLG+RC)	1.00	0.07	0.07
	a. Control by higher Govt.: close local government (CLG)	*No = 1*	*CLG = X*	*1.00*		
	b. Control by higher Govt.: removal of councillors (RC)	*No = 1*	*RC = X*	*1.00*		
20B	Local government authorities (LGA)		LGA = Average (SLT+SYC+BF+CP)	0.67	0.08	0.05

No.	Indicator	Data (X)	Formula	Result	Weight	Total
	c. Local government: set local tax levels (SLT)	*Yes = 1*	*SLT = X*	*1.00*		
	d. Local government: set user charges for services (SUC)	*No = 0*	*SUC = X*	*0.00*		
	e. Local government: borrow funds (BF)	*n.a.*	*BF = X*	*n.a.*		
	f. Local government: choose contractors for projects (CP)	*Yes = 1*	*CP = X*	*1.00*		
21	Codes of conduct (CoC)	Yes = 1	CoC= X	1.00	0.10	0.10
22	Facilities to receive complaints (FRC)		FRC = Average (OA + EF)	1.00	0.10	0.10
	a. Official appointed to receive complaints on public authorities (OA)	*Yes = 1*				
	b. Exclusive facility to receive complaints on corruption (EF)	*Yes = 1*				
23	Anti-corruption commission (ACC)	Yes = 1	ACC = X	1.00	0.15	0.15
24	Personal Income and assets (PIA)		PIA = (0.75* Average PIA + FIA) + 0.25* IAM	0.38	0.15	0.06
	a. Disclosure of personal income and assets (PIA)	*Yes = 1*	*PIA = X*	*1.00*		
	b. Disclosure of family's income and assets (FIA)	*No = 0*	*FIA = X*	*0.00*		
	c. Income and assets regularly monitored (IAM)	*No = 0*	*IAM = X*	*0.00*		
25	Regular independent audit (RIA)	Yes = 1	RIA = X	1.00	0.15	0.15
					Accountability sub-index	**0.85**

"Public Action"

The series "Public Action" studies the State and how it works from a variety of perspectives. It focuses on public policy analysis (environment, health, employment culture etc.) as well as the institutional and organisational study of public administrations. The mutations that characterise nowadays public action (deregulation, externalisation, contractualisation and networking) are at the heart of these preoccupations.

Consequently, the series "Public Action" also places further emphasis on the relationship the State maintains with its environment, – not only the political apparatus and, of course, civil society, but also with broader social changes such as individualism and globalisation.

Multidisciplinary work is an essential characteristic of contemporary research in the field. Given the essential need for such approaches to the topic, this series encourages economic, historical, judicial, politological, sociological and philosophical perspectives.

Series Editors :

Jean-Louis GENARD, Professor at the Université Libre de Bruxelles
Steve JACOB, Professor at the Université Laval

Published Books

N° 1 – Steve JACOB, Frédéric VARONE et Jean-Louis GENARD (dir.), *L'évaluation des politiques au niveau régional*, 2007, 218 p.

N° 2 – Fabrizio CANTELLI, *L'État à tâtons. Pragmatique de l'action publique face au sida*, 2007, 268 p.

No.3 – Isabelle DE LOVINFOSSE, *How and Why Do Policies Change? A Comparison of Renewable Electricity Policies in Belgium, Denmark, Germany, the Netherlands and the UK*, 2008, 317 p.

N° 4 – Frédéric DOBRUSZKES, *Libéralisation et desserte des territoires : le cas du transport aérien européen*, 2008, 287 p.

N° 5 – Fabrizio CANTELLI, Marta ROCA I ESCODA, Joan STAVO-DEBAUGE et Luca PATTARONI (dir.), *Sensibilités pragmatiques. Enquêter sur l'action publique*, 2009, 444 p.

N° 6 – Guy LEBEER & Jacques MORIAU (dir.), *(Se) gouverner. Entre souci de soi et action publique*, 2010, 211 p.

No.7 – Marc LEROY, *Taxation, the State and Society. The Fiscal Sociology of Interventionist Democracy*, 2011, 400 p.

N° 8 – Mathieu BERGER, Daniel CEFAÏ et Carole GAYET-VIAUD (dir.), *Du civil au politique. Ethnographies du vivre-ensemble*, 2011, 603 p.

N° 9 – Thibauld MOULAERT, *Gouverner les fins de carrière à distance.* Outplacement *et vieillissement actif en emploi*, 2012, 287 p.

N° 10 – Léa LIMA (dir.), *L'expertise sur autrui. L'individualisation des politiques sociales entre droit et jugements*, 2013, 242 p.

www.peterlang.com